Problem Books in Mathematics

Series Editor:

Peter Winkler
Department of Mathematics
Dartmouth College
Hanover, NH
USA

More information about this series at http://www.springer.com/series/714

Paul Pollack • Akash Singha Roy

Steps into Analytic Number Theory

A Problem-Based Introduction

Springer

Paul Pollack
Department of Mathematics
University of Georgia
Athens, GA, USA

Akash Singha Roy
Chennai Mathematical Institute
Chennai, Tamil Nadu, India

ISSN 0941-3502   ISSN 2197-8506   (electronic)
Problem Books in Mathematics
ISBN 978-3-030-65079-7   ISBN 978-3-030-65077-3   (eBook)
https://doi.org/10.1007/978-3-030-65077-3

Mathematics Subject Classification: 11-XX, 11-01, 97D50

This Springer imprint is published by the registered company Springer Nature Switzerland AG
The registered company address is: Gewerbestrasse 11, 6330 Cham, Switzerland

For our friends and colleagues at the Ross Mathematics Program.

Preface

These are notes from a 5-week course that I (P.P.) taught as part of the 2019 Ross/Asia Mathematics Program. The course and camp were held from July 7 through August 9 at the Jiangsu Aviation Technical College in Zhenjiang, Jiangsu, China. Clientele ranged widely in age and mathematical experience; most participants were high school students, but some were undergraduates and others were mathematics graduate students.

The course was nontraditional in format. In lieu of lectures, participants received 15 problem sets (or "Steps") over the course of the summer, one every two to three days. Class meetings (held each weekday, and sometimes also on weekends) were entirely devoted to presentations of solutions, sometimes by students and sometimes by the instructor. Participants were expected to have a strong background in mathematical problem solving (e.g., from training for mathematics contests) but not assumed to possess advanced subject-matter knowledge. As such, most of the solutions require nothing more than elementary number theory and a solid grasp of calculus. The chief exception is the proof of Dirichlet's theorem on primes in progressions for a general modulus, where we use (without proof) certain facts from complex variables. Even there, someone familiar with the real-variables side of things will find the necessary results easy enough to swallow.

It is not at all obvious to the uninitiated that analysis has something of value to offer arithmetic. I attempted in this course to marshal the most convincing examples available for this surprising thesis. This explains the somewhat atypical emphasis throughout on concrete, number-theoretic problems, in contrast to a systematic development of analytic tools. Our primary themes are the value distribution of arithmetic functions (e.g., Hardy and Ramanujan's result on the typical number of prime factors of an integer and Erdős's multiplication table theorem), the distribution of prime numbers (Chebyshev's results, Dirichlet's theorem, Brun's theorem on twin primes), and the distribution of squares and nonsquares modulo p (e.g., Vinogradov's upper bound on the least positive nonsquare mod p). Of course, in 5 weeks one can only cover so much; somewhat regrettably, these notes do not include a proof of the Prime Number Theorem.

Apart from the addition of a "problem track" on the values of $\zeta(s)$ at positive even integers, the problem sets are mostly unchanged from what students received in Summer 2019. What *is* new are the solution sets, which have been prepared by myself and Akash Singha Roy (a 2019 Ross Program counselor and enthusiastic participant in the original course). Our intent with the solution sets was to provide enough detail that novices will find the text useful for self-study. Several of the problems come attached to remarks indicating directions for interested students to explore further.

Akash and I would like to conclude by thanking all of the student participants in the original course, as well as the "powers that be" behind the Ross program: Tim All, Jim Fowler, Dan Shapiro, and Jerry Xiao.

Athens, GA, USA   Paul Pollack

Notation

Most of our notation and conventions will already be familiar to anyone who has taken a course in elementary number theory. Here are some possible exceptions: When we write "log", we always intend the natural logarithm. The set of positive integers is denoted $\mathbb{Z}^+$. For us, the letter p always denotes a prime, whether or not this is mentioned explicitly. We use $\mathbb{Z}_m$ for the integers mod m and write $\mathbb{U}_m$ for the group of units mod m. The residue class of the integer a, modulo m, is denoted "a mod m". Finally, sums over integers are to be understood as taken only over positive integers, unless explicitly indicated otherwise. For example, "$\sum_{n \leq x}$" means a sum over positive integers $n \leq x$.

Contents

Step #1

> Do not laugh at notations; invent them, they are powerful. In fact, mathematics is, to a large extent, invention of better notations.
>
> Richard Feynman

Hello to Big-Oh

If f and g are complex-valued functions, we say "f is big-Oh of g", and write $f = O(g)$, to mean that there is a constant $C \geq 0$ such that $|f| \leq C|g|$ for all indicated (or implied) values of the variables. We refer to C as the "implied constant". For instance,

$$x = O(x^2) \quad \text{on} \quad [1, \infty), \qquad \text{with } C = 1 \text{ an acceptable implied constant,}$$

while

$$x \neq O(x^2) \quad \text{on } [0, 1].$$

As a more complicated example,

$$\log(1+x) = x - \frac{1}{2}x^2 + O(x^3) \quad \text{on } [-9/10, 9/10],$$

meaning: there is a function $E(x)$ with $\log(1 + x) = x - \frac{1}{2}x^2 + E(x)$ on $[-9/10, 9/10]$ with $E(x) = O(x^3)$ on $[-9/10, 9/10]$. You can prove this using the Maclaurin series for $\log(1 + x)$. (Really; try it!)

© Springer Nature Switzerland AG 2021
P. Pollack, A. Singha Roy, *Steps into Analytic Number Theory*, Problem Books in Mathematics, https://doi.org/10.1007/978-3-030-65077-3_1

1.1 Basic Properties

(a) For any constant c, we have $c \cdot O(g) = O(g)$.
   Note. Interpret this to mean: "If $f = O(g)$, then $c \cdot f = O(g)$." Parts (b)–(e) should be interpreted similarly.
(b) $O(g) \cdot O(h) = O(gh)$,
(c) $O(f) + O(g) = O(|f| + |g|)$,
(d) If $f = O(g)$ then $O(f) + O(g) = O(g)$,
(e) If $f = O(g)$ and $g = O(h)$, then $f = O(h)$.

1.2 Prove: $\log(1 + x) = x + O(x^2)$ for all $x \geq 0$. Is the same estimate true on $(-0.99, \infty)$? on $(-1, \infty)$?

1.3 We say that $f(x) = O(g(x))$ "as $x \to \infty$" or "for all large x" if $\exists\, x_0$ such that $f(x) = O(g(x))$ on (x_0, ∞). Prove: If $\lim_{x \to \infty} g(x) = 0$, then as $x \to \infty$,

$$\frac{1}{1 + O(g(x))} = 1 + O(g(x)), \quad e^{O(g(x))} = 1 + O(g(x)),$$

$$\text{and} \quad \log(1 + O(g(x))) = O(g(x)).$$

Note. Interpret the first claimed equation to mean that if $f(x) = O(g(x))$ as $x \to \infty$, then $1/(1 + f(x)) = 1 + O(g(x))$, as $x \to \infty$. Similarly for the others.

1.4 As $x \to \infty$,

$$\left(1 + \frac{1}{x}\right)^x = e - \frac{e}{2x} + O\left(\frac{1}{x^2}\right).$$

1.5 If f and g are positive-valued, then $(f + g)^2 \leq 2(f^2 + g^2)$. More generally, for any real $\kappa > 0$, we have $(f + g)^\kappa = O_\kappa(f^\kappa + g^\kappa)$. Here and elsewhere, a subscripted parameter indicates that you are allowed to choose your implied constant to depend on this parameter.

Asymptotic Analysis

1.6 For $n \in \mathbb{Z}^+$, define

$$a_n = \frac{1}{n} - \int_n^{n+1} \frac{dt}{t}.$$

Interpret a_n as an area and explain, from this geometric perspective, how to see that $\sum_{n=1}^{\infty} a_n$ converges.

1.7 There is a real number γ (the "Euler–Mascheroni constant") such that for all positive integers N,

$$0 \geq \sum_{n \leq N} \frac{1}{n} - \left(\log(N+1) + \gamma \right) \geq -\frac{1}{N+1}.$$

1.8 For all real $x \geq 1$: $\displaystyle\sum_{n \leq x} \frac{1}{n} = \log x + \gamma + O(1/x).$

Ingenuity

1.9 (NEWMAN) Let $a_1 = 1$, and let $a_{n+1} = a_n + \frac{1}{a_n}$, for all $n \in \mathbb{Z}^+$. Then $a_n = \sqrt{2n} + O(n^{-1/2} \log n)$, as $n \to \infty$.

Step #2

> Mathematicians have tried in vain to this day to discover some order in the sequence of prime numbers, and we have reason to believe that it is a mystery into which the human mind will never penetrate.
>
> Leonhard Euler

Asymptotic Analysis

If f is strictly decreasing on $[n, n+1]$, then $f(n) > \int_n^{n+1} f(t)\,dt > f(n+1)$ (draw a picture!). If f is strictly increasing, then the inequalities reverse. Use these observations to establish the following estimates.

2.10 For $s > 1$: $\quad \dfrac{1}{s-1} < \sum\limits_{n=1}^{\infty} n^{-s} < \dfrac{s}{s-1}.$

2.11 For $s > 1$ and $x \geq 1$: $\quad \sum\limits_{n>x} n^{-s} < x^{-s} + \dfrac{1}{s-1} x^{1-s} \leq \dfrac{s}{s-1} x^{1-s}.$

2.12 For $x \geq 1$: $\quad \log \lfloor x \rfloor ! = x \log x - x + O(\log (ex)).$ Why do we write ex and not x?

© Springer Nature Switzerland AG 2021
P. Pollack, A. Singha Roy, *Steps into Analytic Number Theory*, Problem Books in
Mathematics, https://doi.org/10.1007/978-3-030-65077-3_2

Infinitely Many Primes

Prove each statement and deduce the infinitude of primes.

2.13 (STIELTJES) If $p_1, \ldots, p_k$ is any finite list of distinct primes, with product P, and ab is any factorization of P into positive integers, then $a + b$ has a prime factor not among $p_1, \ldots, p_k$.

2.14 (GOLDBACH) The "Fermat numbers" $2^{2^n} + 1$, for $n = 0, 1, 2, 3, \ldots$, are pairwise relatively prime.

2.15 (PEROTT) For some constant $c > 0$, and each $N \in \mathbb{Z}^+$, the count of squarefree integers in $[1, N]$ is

$$> N - \sum_{m \geq 2} N/m^2 \geq cN.$$

Thus, there are infinitely many squarefree integers.

2.16 (RAMANUJAN, PILLAI, ENNOLA, RUBINSTEIN) Let $\mathcal{P} = \{p_1, \ldots, p_k\}$ be a set of k primes, where $k < \infty$. For each $x \geq 1$, the number of integers in $[1, x]$ divisible by no primes outside of $\mathcal{P}$ coincides with the number of nonnegative integer solutions $e_1, \ldots, e_k$ to the inequality

$$e_1 \log p_1 + \cdots + e_k \log p_k \leq \log x. \tag{$*$}$$

The number of such solutions is

$$\frac{(\log x)^k}{k! \prod_{i=1}^{k} \log p_i} + O_{\mathcal{P}}((\log (ex))^{k-1}).$$

Hint. Here is a way to start on the upper bound. To each nonnegative integer solution $e_1, \ldots, e_k$ of ($*$), associate the $1 \times 1 \times \cdots \times 1$ (hyper)cube in $\mathbb{R}^k$ having $(e_1, \ldots, e_k)$ as its "leftmost" corner. Show that all of these cubes sit inside the k-dimensional (hyper)tetrahedron defined by '$x_1 \log p_1 + \cdots + x_k \log p_k \leq \log (xp_1 \cdots p_k)$, all $x_i \geq 0$'. What is the volume of that tetrahedron? How does this volume compare to the number of cubes? It might help to first assume that $k = 2$ and draw some pictures.

Combinatorial Methods

2.17 For all $n \in \mathbb{Z}^+$, and all $0 \le r \le n$:

$$\binom{n}{0} - \binom{n}{1} + \cdots + (-1)^r \binom{n}{r} = (-1)^r \binom{n-1}{r}.$$

2.18 For a finite set A, and subsets $A_1, \ldots, A_k$ of A, state and prove the "inclusion-exclusion formula" for $|A \setminus (A_1 \cup A_2 \cup \cdots \cup A_k)|$. Why is it called "inclusion–exclusion"?

2.19 (LEGENDRE)

$$\pi(x) - \pi(\sqrt{x}) + 1$$

$$= \lfloor x \rfloor - \sum_{p_1 \le \sqrt{x}} \left\lfloor \frac{x}{p_1} \right\rfloor + \sum_{p_1 < p_2 \le \sqrt{x}} \left\lfloor \frac{x}{p_1 p_2} \right\rfloor - \sum_{p_1 < p_2 < p_3 \le \sqrt{x}} \left\lfloor \frac{x}{p_1 p_2 p_3} \right\rfloor + \cdots$$

Ingenuity

2.20 (GOLDBACH) If $f(T) \in \mathbb{Z}[T]$ and $f(n)$ is prime for all $n \in \mathbb{Z}^+$, then $f(T)$ is constant.

2.21 (REINER) If k is an integer larger than 1, then the sequence $\{2^{2^n} + k\}_{n=0}^{\infty}$ contains infinitely many composite terms.

Note. It is an open problem to prove this also when $k = 1$.

Step #3

Asymptotic Analysis

The "Euler–Riemann zeta function" $\zeta(s)$ is defined, for $s > 1$, by $\zeta(s) = \sum_{n=1}^{\infty} \frac{1}{n^s}$.

3.22 Justify the "Euler product representation" of the Euler–Riemann zeta function:
For $s > 1$,

$$\zeta(s) = \prod_{p \text{ prime}} \left(1 + \frac{1}{p^s} + \frac{1}{p^{2s}} + \dots\right) = \prod_{p} \frac{1}{1 - \frac{1}{p^s}}.$$

3.23 For $s > 1$: $\log \zeta(s) = \sum_{p} \sum_{k \geq 1} \frac{1}{kp^{ks}} = \sum_{p} \frac{1}{p^s} + O(1).$

3.24 For $1 < s < 2$: $\sum_{p} \frac{1}{p^s} = \log \frac{1}{s-1} + O(1)$. It follows (why?) that $\sum_{p} \frac{1}{p}$
diverges. (EULER)

3.25 Find a sequence $\{c(n)\}_{n=1}^{\infty}$ with the property that

$$\zeta(s) \sum_{n=1}^{\infty} \frac{c(n)}{n^s} = 1$$

© Springer Nature Switzerland AG 2021

P. Pollack, A. Singha Roy, *Steps into Analytic Number Theory*, Problem Books in
Mathematics, https://doi.org/10.1007/978-3-030-65077-3_3

(for all $s > 1$), and describe $c(n)$ in terms of the prime factorization of n. (We will see later that there is a unique such sequence $\{c(n)\}$.)

Combinatorial Methods

3.26 (JORDAN, BONFERRONI) If one halts the inclusion-exclusion formula after an inclusion, one always overshoots (in the sense of obtaining an estimate at least as large as correct). If one stops after an exclusion, one always undershoots.

3.27 Let $\mathcal{A}$ be a set of positive integers. If $\sum_{a \in \mathcal{A}} \frac{1}{a}$ converges, then $\mathcal{A}$ contains 0% of the positive integers, in the sense that

$$\lim_{x \to \infty} \left(\sum_{n \le x, \, n \in \mathcal{A}} 1 \Big/ \sum_{n \le x} 1 \right) = 0.$$

3.28 Let $\mathcal{A}$ be a set of positive integers for which $\sum_{a \in \mathcal{A}} \frac{1}{a}$ diverges. List the elements of $\mathcal{A}$: $a_1 < a_2 < a_3 < \dots$. Then there are infinitely many m for which $a_m < m(\log m)^{1.01}$. It follows that there are arbitrarily large values of x for which

$$\sum_{n \le x, \, n \in \mathcal{A}} 1 > x/(\log x)^{1.01}.$$

Can you think of other functions that can replace $x/(\log x)^{1.01}$ here?

Arithmetic Functions and the Anatomy of Integers

3.29 Suppose that f, g, h are arithmetic functions related by an identity $f(n) = \sum_{d \mid n} g(d)h(n/d)$, valid for all $n \in \mathbb{Z}^+$. Explain why

$$\sum_{n \le x} f(n) = \sum_{a \le x} g(a) \sum_{b \le x/a} h(b) = \sum_{b \le x} h(b) \sum_{a \le x/b} g(a).$$

3.30 For $x \ge 1$: $\sum_{n \le x} \tau(n) = x \log x + O(x)$.   (Thus, a number $n \le x$ has $\approx \log x$ divisors "on average".)

3.31 Large Values of the Divisor Function

(a) The numbers $n = 2^k$ all satisfy $\tau(n) > \log n$.

(b) For every real A, there are infinitely many $n \in \mathbb{Z}^+$ with $\tau(n) > (\log n)^A$.

3.32 For all $n \in \mathbb{Z}^+$: $\tau(n) \leq 2n^{1/2}$.

Ingenuity

3.33 For every $N \in \mathbb{Z}^+$, there is a $d \in \mathbb{Z}^+$ for which the following holds: There are at least N primes p for which $p + d$ is also prime.

Step #4

> What did the analytic number theorist
> say when they were drowning?
> Log-log-log-log-log.
>
> Anonymous

Variations on a Theme of Euler

4.34

(a) For all $x > 0$, and every $\epsilon \in (0, 1)$:

$$\sum_{p \leq x} \frac{1}{p} \leq \sum_{p \leq x} \frac{1}{p}\left(\frac{x}{p}\right)^{\epsilon} = x^{\epsilon} \sum_{p \leq x} \frac{1}{p^{1+\epsilon}} \leq x^{\epsilon} \log \frac{1}{\epsilon} + O(x^{\epsilon}).$$

(b) For all sufficiently large values of x:

$$\sum_{p \leq x} \frac{1}{p} \leq \log \log x + 2 \log \log \log x.$$

Hint. Use (a) with $\epsilon = \frac{1}{\log x \cdot \log \log x}$. (But how did we come up with this choice of ϵ?)

4.35

(a) For all $x > 0$, and every $\epsilon \in (0, 1)$:

© Springer Nature Switzerland AG 2021

P. Pollack, A. Singha Roy, *Steps into Analytic Number Theory*, Problem Books in
Mathematics, https://doi.org/10.1007/978-3-030-65077-3_4

$$\sum_{p \leq x} \frac{1}{p} \geq \sum_{p} \frac{1}{p^{1+\epsilon}} - \sum_{p > x} \frac{1}{p^{1+\epsilon}} \geq \log \frac{1}{\epsilon} - \sum_{n > x} \frac{1}{n^{1+\epsilon}} + O(1).$$

(b) For all sufficiently large values of x:

$$\sum_{p \leq x} \frac{1}{p} \geq \log \log x - 2 \log \log \log x.$$

From Problems 4.34 and 4.35, we conclude that as $x \to \infty$: $\sum_{p \leq x} \frac{1}{p} = \log \log x + O(\log \log \log x)$. Later we will prove sharper estimates for this sum.

Arithmetic Functions and the Anatomy of Integers

4.36 Recall that Euler's ϕ-function satisfies

$$\phi(n) = n \prod_{p \mid n} \left(1 - \frac{1}{p} \right) = n \sum_{d \mid n} \frac{\mu(d)}{d}.$$

Here $\mu(n)$ is the Möbius function, which appeared as the solution sequence $c(n)$ in Problem 3.25. Deduce from Problem 3.29 that for $x \geq 1$:

$$\sum_{n \leq x} \phi(n) = \frac{1}{2} x^2 \sum_{a \leq x} \frac{\mu(a)}{a^2} + O(x \log (ex)).$$

4.37 (DIRICHLET, MERTENS) For $x \geq 1$:

$$\sum_{n \leq x} \phi(n) = \frac{1}{2 \zeta(2)} x^2 + O(x \log (ex)).$$

4.38 (DIRICHLET) A lattice point is chosen uniformly at random from the square $(0, N] \times (0, N]$, where $N \in \mathbb{Z}^+$. As $N \to \infty$, the probability its coordinates are relatively prime tends to $\frac{1}{\zeta(2)}$.

Computing with Roots of Unity

4.39 Let $m \in \mathbb{Z}^+$. For $a \in \mathbb{Z}$:

$$\frac{1}{m} \sum_{k \bmod m} e^{2\pi i k a / m} = \begin{cases} 1 & \text{if } a \equiv 0 \pmod{m}, \\ 0 & \text{otherwise.} \end{cases}$$

Here the sum on k is taken over any set of integer representatives of $\mathbb{Z}_m$.

4.40 (Counting Square Roots mod m) Let $m \in \mathbb{Z}^+$. For $n \in \mathbb{Z}$:

$$\#\{a \bmod m : a^2 \equiv n \ (\bmod \ m)\} = \frac{1}{m} \sum_{k \bmod m} e^{2\pi i k n/m} \sum_{a \bmod m} e^{-2\pi i k a^2/m}.$$

Dirichlet Series

By now we have seen multiple expressions of the form $\sum_{n=1}^{\infty} \frac{a(n)}{n^s}$, where the $a(n)$ are complex numbers. These are known as "Dirichlet series".

4.41 Suppose that $\sum_{n=1}^{\infty} \frac{a(n)}{n^s}$ is a Dirichlet series that converges for some real number $s = s_0$. Then for some real number C, we have $|a(n)| \leq Cn^{s_0}$ for all n. Hence, $\sum_{n=1}^{\infty} \frac{a(n)}{n^s}$ converges absolutely for every $s > s_0 + 1$. Furthermore, for every $m \in \mathbb{Z}^+$:

$$\lim_{s \to \infty} m^s \sum_{n=m}^{\infty} \frac{a(n)}{n^s} = a(m).$$

Mathematical Masterpieces: The Identity as Art Form

4.42 (GOLDBACH) Find the sum of the infinite series

$$\frac{1}{3} + \frac{1}{7} + \frac{1}{8} + \frac{1}{15} + \frac{1}{24} + \frac{1}{26} + \frac{1}{31} + \frac{1}{35} + \frac{1}{48} + \frac{1}{63} + \ldots$$

whose denominators, increased by 1, are the distinct numbers of the form n^m with $n, m \geq 2$ (the perfect powers).

Step #5

> I have had my results for a long time:
> but I do not yet know how I am to
> arrive at them.
>
> ――――――――――――――――――
> Carl Friedrich Gauss

Distribution of Squares mod p

Let p be an odd prime.

5.43 (GAUSS) The "Gauss sum" associated to p is

$$G = \sum_{a \bmod p} e^{2\pi i a^2/p}.$$

Show that for $k \in \mathbb{Z}$, $p \nmid k$:
$$\sum_{a \bmod p} e^{2\pi i k a^2/p} = \left(\frac{k}{p}\right) G.$$

Here $\left(\frac{k}{p}\right)$ is the Legendre symbol: 0 when $p \mid k$, and otherwise 1 or -1, according to whether or not k is a square mod p.

5.44 For $n \in \mathbb{Z}$:

$$\#\{a \bmod p : a^2 \equiv n \pmod{p}\} = 1 + \frac{G}{p} \sum_{k \bmod p} e^{2\pi i k n/p} \left(\frac{-k}{p}\right).$$

Deduce:
$$\left(\frac{n}{p}\right) = \frac{G}{p} \sum_{k \bmod p} e^{2\pi i k n/p} \left(\frac{-k}{p}\right).$$

© Springer Nature Switzerland AG 2021

P. Pollack, A. Singha Roy, *Steps into Analytic Number Theory*, Problem Books in Mathematics, https://doi.org/10.1007/978-3-030-65077-3_5

5.45 Prove: $G \cdot \overline{G} = p$. (Here the bar denotes complex conjugation.) Deduce that G is a square root of $\left(\frac{-1}{p}\right)p$.

[Thus, $G = \pm\sqrt{p}$ when $p \equiv 1 \pmod 4$ and $G = \pm i\sqrt{p}$ when $p \equiv 3 \pmod 4$. Gauss worked for years to determine which sign to take, eventually proving that the $+$ sign is always correct.]

Hint. Start from the expression for $\left(\frac{n}{p}\right)$ proved in Problem 5.44. Take the modulus squared of both sides and sum on n mod p.

Variations on a Theme of Euler

Below, we write $\omega(n)$ for the number of distinct prime factors of n and we use $\Omega(n)$ for the number of prime factors of n, counted with multiplicity. For example, $\omega(45) = 2$, while $\Omega(45) = 3$. Equivalently,

$$\omega(n) = \sum_{p|n} 1, \qquad \Omega(n) = \sum_{p^k|n} 1.$$

5.46 For every nonnegative integer k, and real $x \geq 1$:

$$\sum_{\substack{n \leq x \\ n \text{ squarefree} \\ \omega(n)=k}} \frac{1}{n} \leq \frac{1}{k!} \left(\sum_{p \leq x} \frac{1}{p} \right)^k.$$

5.47 For $x > 1$:

$$\exp\left(\sum_{p \leq x} \frac{1}{p} \right) \geq \sum_{\substack{n \leq x \\ n \text{ squarefree}}} \frac{1}{n}.$$

Also:

$$\zeta(2) \sum_{\substack{n \leq x \\ n \text{ squarefree}}} \frac{1}{n} \geq \sum_{n \leq x} \frac{1}{n} > \log x.$$

Deduce:

$$\sum_{p \leq x} \frac{1}{p} > \log \log x - 1.$$

This improves the lower bound of Problem 4.35.

Arithmetic Functions and the Anatomy of Integers

5.48 (DIRICHLET) For $x \geq 1$: $\displaystyle\sum_{n \leq x} \sigma(n) = \frac{1}{2}\zeta(2)x^2 + O(x \log(ex))$.

5.49 For all $n \in \mathbb{Z}^+$: $2^{\omega(n)} \leq \tau(n) \leq 2^{\Omega(n)}$.

Dirichlet Series

5.50 (KALMÁR) A "multiplicative composition" of n is a representation of n as a product of integers > 1, where order matters. We let $g(n)$ denote the number of multiplicative compositions of n. For instance, $g(1) = 1$ (the empty composition has all parts > 1), while $g(6) = 3$ (for $2 \cdot 3$, $3 \cdot 2$, 6).

n	1	2	3	4	5	6	7	8	9	10	11	12	13	14	15	16	17	18	19	20
$g(n)$	1	1	1	2	1	3	1	4	2	3	1	8	1	3	3	8	1	8	1	8

Let $\rho = 1.72864\ldots$ be the solution in $(1, \infty)$ to $\zeta(\rho) = 2$.
Prove: For all $s > \rho$,

$$\sum_{n=1}^{\infty} \frac{g(n)}{n^s} = \frac{1}{2 - \zeta(s)}.$$

5.51 If $\sum_{n=1}^{\infty} \frac{a(n)}{n^s}$ and $\sum_{n=1}^{\infty} \frac{b(n)}{n^s}$ converge and are equal for all large real numbers s, then each $a(n) = b(n)$. (This implies the uniqueness of the sequence $\{c(n)\}$ in Problem 3.25.)

Mathematical Masterpieces: The Identity as Art Form

5.52 For every nonnegative integer n,

$$\int_0^{\pi/2} \sin^{2n} x \, dx = \frac{\pi}{2} \cdot \frac{1 \cdot 3 \cdots (2n-1)}{2 \cdot 4 \cdots (2n)}, \quad \text{while}$$

$$\int_0^{\pi/2} \sin^{2n+1} x \, dx = \frac{2 \cdot 4 \cdots (2n)}{3 \cdot 5 \cdots (2n+1)}.$$

Here, as usual, empty products are to be understood to equal 1.

5.53 (WALLIS) Show that as $n \to \infty$,

$$\frac{\int_0^{\pi/2} \sin^{2n} x \, dx}{\int_0^{\pi/2} \sin^{2n+1} x \, dx} \to 1.$$

Conclude that $\dfrac{\pi}{2} = \displaystyle\prod_{k=1}^{\infty} \left(\frac{2k}{2k-1} \cdot \frac{2k}{2k+1} \right).$

Step #6

> The elegance of a theorem is directly proportional to the number of ideas you can see in it and inversely proportional to the effort it takes to see them.
>
> George Pólya

Distribution of Squares mod p

Let p be an odd prime.

6.54 Let $k \in \mathbb{Z}$, $p \nmid k$. For all $N \in \mathbb{Z}^+$:

$$\left| \sum_{n \leq N} e^{2\pi i k n / p} \right| = \frac{\left| \sin \frac{\pi k N}{p} \right|}{\left| \sin \frac{\pi k}{p} \right|} \leq \frac{1}{\left| \sin \frac{\pi k}{p} \right|}.$$

6.55 (Pólya–Vinogradov) For all $N \in \mathbb{Z}^+$:

$$\left| \sum_{n \leq N} \left(\frac{n}{p} \right) \right| \leq \sqrt{p} \sum_{n=1}^{(p-1)/2} \frac{1}{n} < \sqrt{p} \log p.$$

Hint. Show (i) $|\sin(\pi\theta)| \geq 2|\theta|$ when $|\theta| \leq \frac{1}{2}$, (ii) $\frac{1}{n} < \int_{-1}^{1} \frac{1}{2n+t} \, dt$.

6.56 Fix $\epsilon \in (0, 1/2)$. For all sufficiently large odd primes p, at least 49% of the integers in $[1, p^{\frac{1}{2}+\epsilon}]$ are squares in $\mathbb{U}_p$, and at least 49% are nonsquares.

In particular, the smallest nonsquare mod p is $< p^{\frac{1}{2}+\epsilon}$.

© Springer Nature Switzerland AG 2021

P. Pollack, A. Singha Roy, *Steps into Analytic Number Theory*, Problem Books in Mathematics, https://doi.org/10.1007/978-3-030-65077-3_6

Combinatorial Methods

6.57 Set $\pi(x, y) = \#\{n \le x : p \mid n \Rightarrow p > y\}$. For $x, y \ge 1$,

$$\pi(x, y) = \lfloor x \rfloor - \sum_{p_1 \le y} \left\lfloor \frac{x}{p_1} \right\rfloor + \sum_{p_1 < p_2 \le y} \left\lfloor \frac{x}{p_1 p_2} \right\rfloor - \sum_{p_1 < p_2 < p_3 \le y} \left\lfloor \frac{x}{p_1 p_2 p_3} \right\rfloor + \dots$$

As a consequence: $\pi(x, y) = x \prod_{p \le y} (1 - 1/p) + O(2^{\pi(y)})$.

6.58 For $y > 1$: $\displaystyle \prod_{p \le y} \left(1 - \frac{1}{p}\right) \le \exp\left(-\sum_{p \le y} \frac{1}{p}\right) = O\left(\frac{1}{\log y}\right)$.

6.59 For $x, y \ge 2$: $\pi(x, y) = O(x/\log y + 2^{\pi(y)})$. Using that $\pi(x) \le \pi(x, y) + y$, and choosing y appropriately, we may deduce that

$$\pi(x) = O(x/\log\log x),$$

as $x \to \infty$. Consequence: 0% of the positive integers are prime.

6.60 (Summation by Parts; ABEL) Let $\{a_n\}_{n=1}^{\infty}$ be a sequence of real or complex numbers, and let $S(t) = \sum_{n \le t} a_n$. If $f(t)$ is any function of t, and $x \ge 1$,

$$\sum_{n \le x} a_n f(n) = S(x) f(\lfloor x \rfloor) + \sum_{n \le x-1} S(n)(f(n) - f(n+1)).$$

6.61 (Continuation) Suppose now that $f(t)$ has a continuous derivative for $t \ge 1$. For each n,

$$S(n)(f(n) - f(n+1)) = -\int_n^{n+1} S(t) f'(t) \, dt.$$

Consequently: For all $x \ge 1$,

$$\sum_{n \le x} a_n f(n) = S(x) f(x) - \int_1^x S(t) f'(t) \, dt.$$

6.62 (CHEBYSHEV) Prove that *if* the ratio $\dfrac{\pi(x)}{x/\log x}$ tends to a limit as $x \to \infty$, then the ratio $\dfrac{\sum_{p \le x} 1/p}{\log\log x}$ tends to the same limit. Conclude that *if* $\dfrac{\pi(x)}{x/\log x}$ tends to a limit, then that limit must be 1.

Hint. Apply summation by parts with $a_n = 1$ if n is prime and 0 otherwise, and $f(t) = 1/t$.

Mathematical Masterpieces: The Identity as Art Form

6.63 Recall that for $|x| \leq 1$,

$$\arcsin(x) = \sum_{n=0}^{\infty} \frac{1 \cdot 3 \cdots (2n-1)}{2 \cdot 4 \cdots 2n} \frac{x^{2n+1}}{2n+1}.$$

[*Reminder of the proof*: Use Newton's binomial theorem to expand $(1 - x^2)^{-1/2}$, then integrate term-by-term.] By plugging in $x = \sin t$ and integrating from 0 to $\frac{\pi}{2}$, deduce that

$$\sum_{n=0}^{\infty} \frac{1}{(2n+1)^2} = \frac{\pi^2}{8} \qquad \text{and that} \qquad \sum_{n=1}^{\infty} \frac{1}{n^2} = \frac{\pi^2}{6}.$$

Ingenuity

6.64 (GARFIELD–KNUTH–WILF) Let $g_e(n)$ and $g_o(n)$ denote the number of multiplicative compositions of n (see Problem 5.50) with an even and an odd number of parts, respectively. Then $g_e(n) - g_o(n)$ is always 0 or ± 1. When does each case occur?

Step #7

Distribution of Prime Numbers

For $x > 0$, define $\mathrm{LCM}(x)$ as the least common multiple of the integers in $[1, x]$. We adopt the convention that $\mathrm{LCM}(x) = 1$ if $0 < x < 1$.

7.65 For $n \in \mathbb{Z}^+$, the ratio $\dfrac{\mathrm{LCM}(2n + 1)}{\mathrm{LCM}(n + 1)}$ is squarefree.

7.66 For $n \in \mathbb{Z}^+$: $\dfrac{\mathrm{LCM}(2n + 1)}{\mathrm{LCM}(n + 1)}$ divides $\dbinom{2n + 1}{n + 1}$.

7.67 For $n \in \mathbb{Z}^+$: $\dbinom{2n + 1}{n + 1} \leq \dfrac{1}{2} \sum_{k \geq 0} \dbinom{2n + 1}{k} = 4^n$.

7.68 $\mathrm{LCM}(n) \leq 4^n$ for all $n \in \mathbb{Z}^+$.

7.69 (GELFOND–SCHNIRELMANN) For $n \in \mathbb{Z}^+$:

$$\frac{1}{4^n} \geq \int_0^1 t^n (1 - t)^n \, dt \in \frac{1}{\mathrm{LCM}(2n + 1)} \mathbb{Z}^+.$$

Hence: $\mathrm{LCM}(2n + 1) \geq 4^n$.

© Springer Nature Switzerland AG 2021
P. Pollack, A. Singha Roy, *Steps into Analytic Number Theory*, Problem Books in Mathematics, https://doi.org/10.1007/978-3-030-65077-3_7

Combinatorial Methods

7.70 Let $m \in \mathbb{Z}^+$ and $x > 0$: The count of $n \le x$ belonging to any given residue class mod m lies within 1 of x/m.

7.71 Set $\pi_2(x, y) = \#\{n \le x : p \mid n(n+2) \Rightarrow p > y\}$. For squarefree d, let $A_d := \#\{n \le x : n(n+2) \equiv 0 \pmod{d}\}$. Then for $x, y \ge 1$,

$$\pi_2(x, y) = A_1 - \sum_{p_1 \le y} A_{p_1} + \sum_{p_1 < p_2 \le y} A_{p_1 p_2} - \sum_{p_1 < p_2 < p_3 \le y} A_{p_1 p_2 p_3} + \cdots$$

$$= \sum_{d \mid P} \mu(d) A_d, \qquad \text{where} \qquad P = \prod_{p \le y} p.$$

7.72 For each squarefree d, we have $A_d = x \frac{2^{\omega(d')}}{d} + O(2^{\omega(d')})$, where d' is the largest odd divisor of d. So for $x, y \ge 2$:

$$\pi_2(x, y) = \frac{1}{2} x \prod_{2 < p \le y} \left(1 - \frac{2}{p}\right) + O(3^{\pi(y)}).$$

What upper bound can you deduce for the "twin prime counting function" $\pi_2(x) := \#\{p \le x : p, p+2 \text{ prime}\}$?

7.73 The usefulness of the above estimate for $\pi_2(x, y)$ is limited by the large error term of $O(3^{\pi(y)})$. To circumvent this, we introduce an alternative approach to bounding $\pi_2(x, y)$ that was suggested by Viggo Brun around 1915.

Let $x, y \ge 1$, and let k be a nonnegative *even* number. Let P be the product of the primes not exceeding y.

(a) $\pi_2(x, y) \le \displaystyle\sum_{\substack{d \mid P, \, \omega(d) \le k}} \mu(d) A_d$.

(b) If we define E_1, E_2 by

$$E_1 = \sum_{\substack{d \mid P \\ \omega(d) > k}} \frac{2^{\omega(d')}}{d}, \qquad E_2 = \sum_{\substack{d \mid P \\ \omega(d) \le k}} 2^{\omega(d')},$$

then

$$\pi_2(x, y) \le \frac{1}{2} x \prod_{2 < p \le y} \left(1 - \frac{2}{p}\right) + x E_1 + E_2.$$

We will return later to the problem of estimating E_1 and E_2.

Mathematical Masterpieces: The Identity as Art Form

7.74 For each integer $k > 1$:

$$\sum_{i=1}^{k-1} \zeta(2i)\zeta(2k-2i)$$

$$= (k-1)\zeta(2k) + \sum_{\substack{m,n \geq 1 \\ m \neq n}} \left(\frac{1}{m^{2k-2}} \frac{1}{n^2 - m^2} + \frac{1}{n^{2k-2}} \frac{1}{m^2 - n^2} \right).$$

Ingenuity

7.75 (NAIR) Explicitly evaluate $\int_0^1 t^n (1-t)^n \, dt$, for $n \in \mathbb{Z}^+$. Use this to show that $LCM(2n+1) \geq 2^{2n+2}$ for all large enough n, and conclude that $LCM(n) \geq 2^n$ for all large n.

Step #8

Distribution of Squares mod p

8.76 Let p be an odd prime. Let $r_1 < r_2 < \ldots < r_{(p-1)/2}$ and $n_1 < n_2 < \ldots < n_{(p-1)/2}$ be the lists of integers in $(0, p)$ that represent squares and nonsquares in $\mathbb{U}_p$. For each $\epsilon > 0$, and all large p, each $r_{i+1} - r_i < p^{\frac{1}{2}+\epsilon}$, and each $n_{i+1} - n_i < p^{\frac{1}{2}+\epsilon}$.

Distribution of Prime Numbers

8.77 For all $x > 0$: $\quad \pi(x)! \leq \prod_{p \leq x} p \leq \mathrm{LCM}(x)$.

8.78 (CHEBYSHEV) Fix $\epsilon > 0$. For all sufficiently large x:

$$\pi(x) \leq (\log 4 + \epsilon) \frac{x}{\log x}.$$

8.79 For $x > 0$: $\quad \log \mathrm{LCM}(x) = \sum_{p \leq x} \log p \left\lfloor \frac{\log x}{\log p} \right\rfloor \leq \pi(x) \log x$.

8.80 (CHEBYSHEV) Fix $\epsilon > 0$. For all sufficiently large x:

© Springer Nature Switzerland AG 2021
P. Pollack, A. Singha Roy, *Steps into Analytic Number Theory*, Problem Books in
Mathematics, https://doi.org/10.1007/978-3-030-65077-3_8

$$\pi(x) \geq (\log 2 - \epsilon)\frac{x}{\log x}.$$

8.81 Fix $K > 2$. For all sufficiently large x, there is a prime in the interval $(x, Kx]$.

8.82 For $x > 0$: $\lfloor x \rfloor! = \prod_{m \leq x} \mathrm{LCM}(x/m)$.

Variations on a Theme of Euler

8.83 Let $\chi : \mathbb{Z} \to \{-1, 0, 1\}$ be defined by

$$\chi(n) = \begin{cases} 1 & \text{if } n \equiv 1 \pmod 3, \\ -1 & \text{if } n \equiv -1 \pmod 3, \\ 0 & \text{if } 3 \mid n. \end{cases}$$

Put $\qquad L(s) = \sum_{n=1}^{\infty} \frac{\chi(n)}{n^s}$

$$= 1 - \frac{1}{2^s} + \frac{1}{4^s} - \frac{1}{5^s} + \frac{1}{7^s} - \frac{1}{8^s} + \frac{1}{10^s} - \frac{1}{11^s} + \dots.$$

Show that for $s > 1$:

(a) $L(s) = \prod_p \dfrac{1}{1 - \frac{\chi(p)}{p^s}}$.

(b) $1 \geq L(s) \geq 1/2$. As a consequence, $\log L(s) = O(1)$.

(c) $\displaystyle\sum_{p \equiv 1 \pmod 3} \frac{1}{p^s} - \sum_{p \equiv -1 \pmod 3} \frac{1}{p^s} = O(1)$.

(d) For $1 < s < 2$, and either choice of $\pm$ sign:

$$\sum_{p \equiv \pm 1 \pmod 3} \frac{1}{p^s} = \frac{1}{2} \log \frac{1}{s-1} + O(1).$$

8.84 Both residue classes 1 mod 3 and -1 mod 3 contain infinitely many primes. In fact, the sum of the reciprocals of the primes up to x in either progression is $\frac{1}{2} \log \log x + O(\log \log \log x)$, as $x \to \infty$.

8.85 The conclusion of Problem 8.84 also holds for the residue classes 1 mod 4 and -1 mod 4.

Ingenuity

8.86 (V.-A. LEBESGUE) Let p be an odd prime. For each positive integer ℓ, let $S_\ell = \#\{(x_1, \dots, x_\ell) \in (\mathbb{Z}_p)^\ell : x_1^2 + \cdots + x_\ell^2 = 1\}$. By writing

$$S_\ell = \frac{1}{p} \sum_{x_1, \dots, x_\ell \bmod p} \sum_{k \bmod p} e^{2\pi i k(x_1^2 + \cdots + x_\ell^2 - 1)/p}$$

and using your evaluation of the Gauss sums, find a simple, closed form expression for S_ℓ, for each ℓ.

Step #9

When I encountered Brun's sieve for the first time, I was reminded of the legend that Alexander the Great cut with his sword the intricate knot of Phrygian King Gordius, and proceeded to Asia. In fact, in my mind Brun is mightier than the great king, for he cut the enigmatic knot that had survived 2100 years without any sign of wear.

Yoichi Motohashi

Arithmetic Functions and the Anatomy of Integers

9.87 For $n \in \mathbb{Z}^+$: $|\mu(n)| = \sum_{d^2|n} \mu(d)$. The sum here is over all $d \in \mathbb{Z}^+$ for which $d^2 \mid n$.

9.88 (GEGENBAUER) For $x \geq 1$:

$$\sum_{n \leq x} |\mu(n)| = \sum_{d \leq \sqrt{x}} \mu(d) \left\lfloor \frac{x}{d^2} \right\rfloor = \frac{1}{\zeta(2)} x + O(x^{1/2}).$$

Hence, a "random" integer is squarefree with probability $\zeta(2)^{-1}$. Can you generalize to kth-power-free integers?

© Springer Nature Switzerland AG 2021
P. Pollack, A. Singha Roy, *Steps into Analytic Number Theory*, Problem Books in Mathematics, https://doi.org/10.1007/978-3-030-65077-3_9

Combinatorial Methods

We recall some of the setup of Step #7. Let

$$\pi_2(x, y) = \#\{n \leq x : p \mid n(n+2) \Rightarrow p > y\}.$$

Let $x, y \geq 2$, and let P be the product of the primes not exceeding y. You have proved (Problem 7.73) that for every nonnegative even integer k,

$$\pi_2(x, y) \leq \frac{1}{2}x \prod_{2 < p \leq y} \left(1 - \frac{2}{p}\right) + xE_1 + E_2,$$

where $E_1 = \sum_{\substack{d \mid P \\ \omega(d) > k}} \frac{2^{\omega(d')}}{d}$, and $E_2 = \sum_{\substack{d \mid P \\ \omega(d) \leq k}} 2^{\omega(d')}$. Here d' is the largest odd divisor of d.

Assume in the problems below that $k = k(x)$ and $y = y(x)$ are parameters depending on x, both of which tend to infinity with x. (We wait until Problem 9.91 to pin down k, y explicitly.)

9.89 For large enough x:

$$E_1 \leq \sum_{\substack{d \mid P \\ \omega(d) > k}} \frac{2^{\omega(d)}}{d} \leq \sum_{d \mid P} \frac{2^{\omega(d)}}{d} e^{\omega(d) - k} = e^{-k} \prod_{p \leq y} \left(1 + \frac{2e}{p}\right) \leq e^{-k}(\log y)^6.$$

9.90 For large enough x: Every $d \mid P$ with $\omega(d) \leq k$ satisfies $d \leq y^k$, and so $E_2 \leq y^k \max_{d \leq y^k} 2^{\omega(d)} \leq y^{2k}$.

9.91 Now choose $y = x^{\frac{1}{100 \log\log x}}$ and $k = 10\lfloor \log\log x \rfloor$. Deduce:

$$\pi_2(x, x^{\frac{1}{100 \log\log x}}) = O(x(\log\log x)^2/(\log x)^2),$$

as $x \to \infty$. Conclude: $\pi_2(x) = O(x(\log\log x)^2/(\log x)^2)$, as $x \to \infty$.

9.92 (BRUN) If $\mathcal{P}$ is the set of primes p for which $p + 2$ is also prime, then $\sum_{p \in \mathcal{P}} \frac{1}{p} < \infty$.

Distribution of Squares mod p

9.93 (WHYBURN) Let p be an odd prime. The reduced fractions a/b, with $0 < a, b < \sqrt{p}$, are all distinct in $\mathbb{Z}_p$. Deduce: For large p, these fractions represent more than 50% of the nonzero elements of $\mathbb{Z}_p$.

9.94 For all large primes p, the least positive nonsquare modulo p is smaller than $\sqrt{p}$.

Mathematical Masterpieces: The Identity as Art Form

9.95 Evaluate $\displaystyle\sum_{n>1} \frac{1}{n^2 - 1}$. More generally, given $m \in \mathbb{Z}^+$, determine a simple closed form expression for $\displaystyle\sum_{\substack{n \in \mathbb{Z}^+ \\ n \neq m}} \frac{1}{n^2 - m^2}$.

Ingenuity

9.96 For integers $n > 1$, write $P(n)$ for the largest prime factor of n. For which real numbers λ does the series $\sum_{n>1} \frac{1}{n^\lambda P(n)}$ converge ?

Step #10

A mathematician is a conjurer who gives away [their] secrets.

John H. Conway

Variations on a Theme of Euler

10.97

(a) Using Legendre's formula for the prime factorization of $\lfloor x \rfloor!$, deduce that for $x > 0$:

$$\log \lfloor x \rfloor! = x \sum_{p \le x} \frac{\log p}{p} + O(x).$$

(b) For $x \ge 1$: $\displaystyle\sum_{p \le x} \frac{\log p}{p} = \log x + O(1).$

10.98 (MERTENS) Write $\sum_{p \le x} \frac{\log p}{p} = \log x + E(x)$, so that $E(x) = O(1)$ by Problem 10.97(b). Show that for $x \ge 2$:

$$\sum_{p \le x} \frac{1}{p} = \log \log x + C + O\left(\frac{1}{\log x}\right),$$

where the constant C is given by $C = 1 - \log \log 2 + \int_2^\infty \frac{E(t)}{t(\log t)^2}\, dt.$

© Springer Nature Switzerland AG 2021
P. Pollack, A. Singha Roy, *Steps into Analytic Number Theory*, Problem Books in Mathematics, https://doi.org/10.1007/978-3-030-65077-3_10

Arithmetic Functions and the Anatomy of Integers

10.99

(a) The proportion of numbers in $[1, x]$ possessing a prime factor $> \sqrt{x}$ tends to $\log 2$, as $x \to \infty$.

(b) If we would like the proportion of numbers in $[1, x]$ with a prime factor $> y$ to tend to $\frac{1}{2}$, as $x \to \infty$, how could we choose $y = y(x)$?

10.100 Mean values of $\omega(n)$ and $\omega(n)^2$

(a) As $x \to \infty$: $\displaystyle\sum_{n \le x} \omega(n) = x \log\log x + O(x)$. (Thus, $\omega(n) \approx \log\log x$ "on average" over $n \le x$.)

(b) $\displaystyle\sum_{n \le x} \omega(n)^2 = x \sum_{\substack{p,q \text{ prime} \\ pq \le x}} \frac{1}{pq} + O(x \log\log x)$.   *Hint:* $\omega(n)^2 = \sum_{\substack{p,q \text{ prime} \\ p,q \mid n}} 1$.

(c) As $x \to \infty$: $\displaystyle\sum_{n \le x} \omega(n)^2 = x(\log\log x)^2 + O(x \log\log x)$.

10.101 For $x > 0$: $\sum_{n \le x}(\Omega(n) - \omega(n)) = O(x)$. Thus, $\Omega(n)$ and $\omega(n)$ differ by a bounded amount on average.

10.102 For all large integers n: $n/\phi(n) = O(\log\log n)$. *Hint:* First show $\prod_{p \mid n,\ p > \log n}(1 - 1/p)^{-1} = O(1)$.

Order of 2 mod ℓ

Let $n > 1$, and let $B_n = \gcd\left\{ \dfrac{2^n - 1}{2^{n/p} - 1} : p \mid n \right\}$.

n	2	3	4	5	6	7	8	9	10	11	12	13	14	15	16	17	18
B_n	3	7	5	31	3	127	17	73	11	$23 \cdot 89$	13	8191	43	151	257	131,071	$3 \cdot 19$

10.103 Certainly if ℓ is a prime for which $o(2 \bmod \ell) = n$, then $\ell \mid B_n$. (Make sure you see why!) We now investigate the converse.

Suppose $\ell \mid B_n$, with ℓ prime. Since $\ell \mid 2^n - 1$, we know $o(2 \bmod \ell) \mid n$. Assuming that $o(2 \bmod \ell) < n$, show:

(a) $o(2 \bmod \ell)$ divides n/p for some prime p dividing n,

(b) The unique prime p satisfying (a) is $p = \ell$. As a consequence, $\ell \mid n$.

(c) If q is a prime dividing n with $q \ne \ell$, then q divides $o(2 \bmod \ell)$.

(d) ℓ is the largest prime factor of n.

(e) $\ell^2 \nmid B_n$.

10.104 If there does not exist a prime number ℓ for which $o(2 \bmod \ell) = n$, call n "repellent". Show that if n is repellent, then $B_n = 1$ or B_n is the largest prime factor of n.

Do you have a guess as to which n are repellent?

Mathematical Masterpieces: The Identity as Art Form

10.105 (EULER) For each integer $k > 1$:

$$\sum_{i=1}^{k-1} \zeta(2i)\zeta(2k - 2i) = \left(k + \frac{1}{2}\right)\zeta(2k).$$

Thus, $\zeta(4) = \dfrac{\pi^4}{90}$, and in general, $\zeta(2k) \in \pi^{2k}\mathbb{Q}$.

Ingenuity

10.106 Recall that a "perfect number" is a positive integer n for which $n = \sum_{d|n,\, d<n} d$.

(a) (DESCARTES) Every odd perfect number can be written in the form $p^k m^2$, where p is prime, $k \in \mathbb{Z}^+$, and $p \nmid m$.

(b) (HORNFECK) Use (a) to prove that for all $x > 0$, the number of odd perfect numbers in $[1, x]$ is $\le x^{1/2}$.

Step #11

> There are many things you can do with problems besides solving them. First you must define them, pose them. But then of course you can also refine them, depose them, or repose them, even dissolve them!
>
> David Hawkins

Distribution of Squares mod p

11.107 Fix $\epsilon \in (0, \frac{1}{50})$. For all large primes p, fewer than 45% of the integers in $[1, p^{\frac{1}{2}+\epsilon}]$ have a prime factor $> p^{1/3}$.

11.108 For all large primes p, the smallest positive nonsquare modulo p is smaller than $p^{1/3}$.

11.109 (VINOGRADOV) The conclusion of Problem 11.108 holds with $\frac{1}{3}$ replaced by any real number $> \frac{1}{2\sqrt{e}}$.

Arithmetic Functions and the Anatomy of Integers

11.110 (TURÁN) As $x \to \infty$: $\displaystyle\sum_{n \le x}(\omega(n) - \log\log x)^2 = O(x \log\log x)$.

© Springer Nature Switzerland AG 2021
P. Pollack, A. Singha Roy, *Steps into Analytic Number Theory*, Problem Books in Mathematics, https://doi.org/10.1007/978-3-030-65077-3_11

11.111 (HARDY–RAMANUJAN) For large x, we have $\omega(n) \approx \log\log x$ for "most" values of $n \le x$. Precisely: Fix $\epsilon > 0$. Then the proportion of $n \le x$ for which $|\omega(n) - \log\log x| > \epsilon \log\log x$ tends to 0, as $x \to \infty$.

11.112 (HARDY–RAMANUJAN) The conclusion of Problem 11.111 holds with $\Omega(n)$ replacing $\omega(n)$.

11.113 (HARDY–RAMANUJAN) For large x, we have $\tau(n) \approx (\log x)^{\log 2}$ for "most" values of $n \le x$. Precisely: Fix $\epsilon > 0$. Then the proportion of $n \le x$ for which $\tau(n) \notin ((\log x)^{\log 2 - \epsilon}, (\log x)^{\log 2 + \epsilon})$ tends to 0, as $x \to \infty$.

How does $(\log x)^{\log 2}$ compare with the average of $\tau(n)$, of size $\approx \log x$, determined in Problem 3.30?

11.114 (Maximal Size of $\tau(n)$; WIGERT) Given a large positive integer n, factor $n = AB$, where A is the largest divisor of n composed of primes not exceeding $\log n/(\log\log n)^3$. Prove:

(a) $\tau(A) \le 2^{O(\log n/(\log\log n)^2)}$.

(b) $\Omega(B) \le \frac{\log n}{\log\log n} + O\left(\frac{\log n \cdot \log\log\log n}{(\log\log n)^2}\right)$, and so

$$\tau(B) \le 2^{\frac{\log n}{\log\log n} + O\left(\frac{\log n \cdot \log\log\log n}{(\log\log n)^2}\right)}.$$

(c) For each $\epsilon > 0$ and all sufficiently large n:

$$\tau(n) \le 2^{(1+\epsilon)\log n/\log\log n}.$$

(d) For each $\epsilon > 0$, there are infinitely many n such that

$$\tau(n) \ge 2^{(1-\epsilon)\log n/\log\log n}.$$

In fact, that inequality holds for $n = \mathrm{LCM}(x)$, once x is sufficiently large.

Hint. $\tau(\mathrm{LCM}(x)) \ge 2^{\pi(x)}$. Find a way to use the result of Exercise 8.79.

Note. Since $2^{(1+\epsilon)\log n/\log\log n} = n^{(1+\epsilon)\log 2/\log\log n}$, and $1/\log\log n \to 0$, part (c) has as a weak consequence that $\tau(n) = O_\epsilon(n^\epsilon)$ for all $\epsilon > 0$.

Order of 2 mod ℓ

Recall the definition of the "cyclotomic polynomials" $\Phi_n(T)$: $\Phi_1(T) = T - 1$ and successive values of $\Phi_n(T)$ are determined by the relation

$$\prod_{d|n} \Phi_d(T) = T^n - 1, \qquad \text{for all } n \in \mathbb{Z}^+. \tag{*}$$

Here are the first several values of $\Phi_n(T)$:

n	1	2	3	4	5	6
$\Phi_n(T)$	$T-1$	$T+1$	T^2+T+1	T^2+1	$T^4+T^3+T^2+T+1$	T^2-T+1

A priori, each $\Phi_n(T) \in \mathbb{Q}(T)$, but an induction argument shows that each $\Phi_n(T) \in \mathbb{Z}[T]$. From (*) and Möbius inversion,

$$\Phi_n(T) = \prod_{d \mid n} (T^d - 1)^{\mu(n/d)}.$$

11.115 For each $n > 1$, the integer $\Phi_n(2)$ divides the number B_n defined in Step #10.

11.116 For all $n \in \mathbb{Z}^+$: $\quad \Phi_n(2) = 2^{\phi(n)} \prod_{d \mid n} (1 - 1/2^d)^{\mu(n/d)} > \frac{1}{3} 2^{\phi(n)}$.

11.117 (BANG) For every $n > 6$, there is a prime ℓ with $o(2 \bmod \ell) = n$.

Step #12

> I like to think of mathematicians as
> forming a nation of our own without
> distinctions of geographical origin,
> race, creed, sex, age or even time... all
> dedicated to the most beautiful of the
> arts and sciences.
>
> Julia Robinson

Arithmetic Functions and the Anatomy of Integers

For $N \in \mathbb{Z}^+$, let $\mathcal{M}(N) = \{m \cdot n : 1 \le m, n \le N\}$ be the collection of numbers appearing in the $N \times N$ multiplication table, and put $M(N) = \#\mathcal{M}(N)$.

N	1	2	4	8	16	32	64	128	256	512	1024
$M(N)$	1	3	9	30	97	354	1263	4695	17,668	67,765	260,095
$M(N)/N^2$	1.00	0.75	0.56	0.47	0.38	0.35	0.31	0.29	0.27	0.26	0.25

12.118 Let $\epsilon > 0$. If $m, n \in \mathbb{Z}^+$ are selected uniformly at random from $[1, N]$, independently, then the probability that $\Omega(mn)$ is more than $\epsilon \log \log N$ away from $2 \log \log N$ tends to 0, as $N \to \infty$.

12.119 Let $\epsilon > 0$. If $n \in \mathbb{Z}^+$ is selected uniformly at random from $[1, N^2]$, then the probability that $\Omega(n)$ is more than $\epsilon \log \log N$ away from $\log \log N$ tends to 0, as $N \to \infty$.

12.120 (ERDŐS) $\displaystyle\lim_{N \to \infty} \frac{M(N)}{N^2} = 0$.

© Springer Nature Switzerland AG 2021
P. Pollack, A. Singha Roy, *Steps into Analytic Number Theory*, Problem Books in
Mathematics, https://doi.org/10.1007/978-3-030-65077-3_12

Distribution of Prime Numbers

12.121 (RAMANUJAN) For $n \in \mathbb{Z}^+$:

(a) $\dfrac{\text{LCM}(2n) \cdot \text{LCM}(2n/3) \cdot \text{LCM}(2n/5) \cdots}{\text{LCM}(n) \cdot \text{LCM}(n/2) \cdot \text{LCM}(n/3) \cdots} = \dbinom{2n}{n} \geq \dfrac{4^n}{2n}.$

(b) $\dfrac{\text{LCM}(2n)}{\text{LCM}(n)} \text{LCM}(2n/3) \geq \dfrac{4^n}{2n}.$

(c) $\dfrac{\text{LCM}(2n)}{\text{LCM}(n)} \geq \dfrac{4^{n/3}}{2n}.$

(d) $\dfrac{\text{LCM}(2n)}{\text{LCM}(n)}$ is squarefree, with every prime divisor belonging to $(n, 2n]$ or at most $\sqrt{2n}$.

(e) There is always a prime between n and $2n$—meaning, in the interval $(n, 2n]$.

12.122 (PILLAI) For every $n > 1$, there is a prime congruent to 1 mod n that is $< 2^n$.

At the end of the nineteenth century, Hadamard and de la Vallée Poussin (independently) established the "Prime Number Theorem" (PNT):

$$\frac{\pi(x)}{x/\log x} \to 1, \qquad \text{as } x \to \infty.$$

Assume the PNT for the next two problems.

12.123 For every finite sequence $d_1, \ldots, d_n \in \{0, 1, 2, \ldots, 9\}$, with $d_1 \neq 0$, there is a prime number whose leftmost n decimal digits are $d_1, \ldots, d_n$.

12.124

(a) Show that $\lim\limits_{x \to \infty} \dfrac{\sum_{p \leq x} P}{x^2/\log x} = \dfrac{1}{2}.$

(b) (ALLADI) Let $A(n) = \sum_{p|n} p$. Determine $\lim\limits_{x \to \infty} \dfrac{\sum_{1 \leq n \leq x} A(n)}{x^2/\log x}.$

Ingenuity

12.125 (PENNEY–POMERANCE) Let $S_k(n) = 1^k + 2^k + 3^k + \cdots + n^k$. It is well-known that $S_3(n) = S_1(n) \cdot S_1(n)$ for all $n \in \mathbb{Z}^+$ (both sides $= (\frac{1}{2}n(n+1))^2$). Classify all (nonempty) multisets of positive integers $\{a_1, \ldots, a_k\}, \{b_1, \ldots, b_\ell\}$ for which

$$\prod_i S_{a_i}(n) = \prod_j S_{b_j}(n) \quad \text{for all } n \in \mathbb{Z}^+.$$

Special Step A: Dirichlet's Theorem for $m = 8$

> Dirichlet alone, not I, nor Cauchy, nor Gauss knows what a completely rigorous mathematical proof is. Rather we learn it first from him. When Gauss says that he proved something, it seems to me very probable, when Cauchy says it, you can wager as much pro as con; when Dirichlet says it, it is certain.
>
> Carl Gustav Jacob Jacobi

In the 1830s, Dirichlet proved that there are infinitely many primes $p \equiv a$ (mod m) whenever a and m are relatively prime.

Our goal on this set is to show, by Dirichlet's method, that there are infinitely many primes in each of the residue classes 1 mod 8, 3 mod 8, 5 mod 8, and 7 mod 8. This proof of Dirichlet's theorem for $m = 8$ occupies a middle ground between the relatively simple arguments of Step #8 for the cases $m = 3$ and $m = 4$, and the more intricate arguments of the next two Special Steps, where Dirichlet's theorem is proved in general.

Characters of $\mathbb{U}_8$

If G is a finite commutative group, by a "character" of G we mean a homomorphism $\chi : G \rightarrow \{z \in \mathbb{C} : |z| = 1\}$ (the target set here forming a group under multiplication). We use the symbol χ_0 to denote the "trivial character" mapping all

© Springer Nature Switzerland AG 2021
P. Pollack, A. Singha Roy, *Steps into Analytic Number Theory*, Problem Books in Mathematics, https://doi.org/10.1007/978-3-030-65077-3_13

Table 13.1 Character table
of $\mathbb{U}_3$

	χ_0	χ
1 mod 3	1	1
2 mod 3	1	−1

of G to 1. For example, when $G = \mathbb{U}_3$ there is precisely one nontrivial character χ, and the table of characters along with their values is very simple (shown at right, below) (Table 13.1).

13.126 Show that if χ is a character of $\mathbb{U}_8$, then $\chi(g)^2 = 1$ for all $g \in \mathbb{U}_8$. Use this to show that $\mathbb{U}_8$ has exactly 3 nontrivial characters, with character table as follows:

	χ_0	χ_1	χ_2	χ_3
1 mod 8	1	1	1	1
3 mod 8	1	−1	−1	1
5 mod 8	1	1	−1	−1
7 mod 8	1	−1	1	−1

13.127 Calculate the sum of each row of the table. Conclude that the indicator function $\mathcal{I}_{1 \bmod 8} \colon \mathbb{U}_8 \to \{0, 1\}$ corresponding to the element $1 \bmod 8 \in \mathbb{U}_8$ is given by $\mathcal{I}_{1 \bmod 8} = \frac{1}{4}(\chi_0 + \chi_1 + \chi_2 + \chi_3)$.

13.128 Let a be an odd integer. Using that $n \bmod 8 = a \bmod 8 \iff (n \bmod 8)(a \bmod 8)^{-1} = 1 \bmod 8$, deduce that

$$\mathcal{I}_{a \bmod 8} = \frac{1}{4} \sum_{\chi} \chi(a \bmod 8)^{-1} \chi,$$

where the sum on χ is over all the characters of $\mathbb{U}_8$. Conclude:

$$\mathcal{I}_{1 \bmod 8} = \frac{1}{4}(\chi_0 + \chi_1 + \chi_2 + \chi_3), \quad \mathcal{I}_{3 \bmod 8} = \frac{1}{4}(\chi_0 - \chi_1 - \chi_2 + \chi_3),$$

$$\mathcal{I}_{5 \bmod 8} = \frac{1}{4}(\chi_0 + \chi_1 - \chi_2 - \chi_3), \quad \mathcal{I}_{7 \bmod 8} = \frac{1}{4}(\chi_0 - \chi_1 + \chi_2 - \chi_3).$$

Dirichlet L-Functions for $m = 8$

Let m be a positive integer. If χ is a character of $\mathbb{U}_m$, we associate to χ the function on $\mathbb{Z}$ which takes the value $\chi(n \bmod m)$ at integers n coprime to m, and the value 0 at all other integers. We use the same notation for both the original function on $\mathbb{U}_m$ and this "lift" to a function on $\mathbb{Z}$; this will not cause any confusion.

The functions on $\mathbb{Z}$ that arise this way are called the "Dirichlet characters" modulo m. By construction, each Dirichlet character χ satisfies

$$\chi(ab) = \chi(a)\chi(b) \qquad \text{for all } a, b \in \mathbb{Z}.$$

For each Dirichlet character χ, we define the "Dirichlet L-function" associated to χ by

$$L(s, \chi) = \sum_{n=1}^{\infty} \frac{\chi(n)}{n^s}.$$

The unique nontrivial character χ mod 3, along with its associated L-function, appeared already in Problem 8.84 (there $L(s, \chi)$ was denoted $L(s)$).

13.129 Let χ be any Dirichlet character modulo a positive integer m. Show that the series defining $L(s, \chi)$ is absolutely convergent in the complex half-plane $\Re(s) > 1$. Deduce that for $\Re(s) > 1$,

$$L(s, \chi) = \prod_{p} \frac{1}{1 - \frac{\chi(p)}{p^s}}.$$

This generalizes Problem 8.84(a).

For the rest of the problem set, we retreat from general Dirichlet characters to the safety of $m = 8$.

13.130 Let χ be a Dirichlet character mod 8. Show that for all $s > 1$, the double series $\sum_p \sum_{k \geq 1} \frac{\chi(p^k)}{kp^{ks}}$ converges absolutely (so that the terms may be safely reordered), and

$$\exp\left(\sum_p \sum_{k \geq 1} \frac{\chi(p^k)}{kp^{ks}}\right) = L(s, \chi).$$

Deduce that $L(s, \chi)$ is positive and that $\log L(s, \chi) = \sum_{\substack{p \text{ prime} \\ k \geq 1}} \frac{\chi(p^k)}{kp^{ks}}$.

13.131 Let χ be a nontrivial Dirichlet character mod 8. You will show in this problem that the nonvanishing of $L(1, \chi)$ implies that $\log L(s, \chi)$ is bounded for s near 1. Set

$$A_\chi(t) = \sum_{n \leq t} \chi(n).$$

(a) Prove that

$$\sum_{n \le x} \chi(n) n^{-s} = x^{-s} A_\chi(x) + s \int_1^x \frac{A_\chi(t)}{t^{s+1}} \, dt$$

for all $x \ge 1$.

(b) By examining the column sums of the character table, show that $|A_\chi(t)| \le 2$ for all $t \ge 0$.

(c) Deduce from (a), (b) that the series defining $L(s, \chi)$ converges for all $s > 0$, to $s \int_1^\infty \frac{A_\chi(t)}{t^{s+1}} \, dt$.

(d) Use these integral representations to prove that the series defining $L(s, \chi)$ converges *uniformly* for $s \ge \epsilon$, for any fixed $\epsilon > 0$. Since the summands are continuous functions of s, a standard theorem in real analysis guarantees that $L(s, \chi)$ is continuous for $s > 0$.

(e) Deduce: If $L(1, \chi) \ne 0$, then $\log L(s, \chi)$ is defined and continuous for all $s \ge 1$. Conclude that $\log L(s, \chi) = O(1)$ for $1 < s < 2$.

Reduction to the Nonvanishing of $L(1, \chi)$ for $\chi \ne \chi_0$

It follows from Problem 13.128 that for every odd integer a, and all $s > 1$,

$$\sum_{p \equiv a \pmod 8} \frac{1}{p^s} = \sum_{p \text{ odd}} \frac{1}{p^s} \cdot \mathcal{I}_{a \bmod 8}(p \bmod 8)$$

$$= \sum_{p \text{ odd}} \frac{1}{p^s} \left(\frac{1}{4} \sum_\chi \chi(a)^{-1} \chi(p) \right) = \frac{1}{4} \sum_\chi \chi(a)^{-1} \sum_p \frac{\chi(p)}{p^s},$$

where the sums on χ are over all Dirichlet characters χ mod 8.

13.132 Estimating $\sum_p \chi(p) p^{-s}$

(a) For each Dirichlet character χ mod 8:

$$\sum_p \frac{\chi(p)}{p^s} = \log L(s, \chi) + O(1) \quad \text{for } 1 < s < 2.$$

(b) $L(s, \chi_0) = (1 - 2^{-s}) \zeta(s)$ when $s > 1$. Hence:

$$\sum_p \frac{\chi_0(p)}{p^s} = \log \frac{1}{s - 1} + O(1) \quad \text{for } 1 < s < 2.$$

(c) Suppose $\chi \neq \chi_0$ and $L(1, \chi) \neq 0$. Then

$$\sum_p \frac{\chi(p)}{p^s} = O(1) \quad \text{for } 1 < s < 2.$$

13.133 Assume $L(1, \chi) \neq 0$ for all $\chi \neq \chi_0$. Then for every odd a, and all s with $1 < s < 2$,

$$\sum_{p \equiv a \pmod 8} \frac{1}{p^s} = \frac{1}{4} \log \frac{1}{s-1} + O(1).$$

Consequently, there are infinitely many primes $p \equiv a \pmod 8$.

Nonvanishing of $L(1, \chi)$ for $\chi \neq \chi_0$

13.134 Write out the series for $L(1, \chi)$, for each nontrivial character χ mod 8. Use familiar facts about alternating series to prove that each such series converges to a positive real number.

Techniques of Generalization

13.135 Use the ideas of this problem set to prove Dirichlet's theorem when $m = 12$.

Special Step B: Dirichlet's Theorem for $m = \ell$ (Odd Prime)

In this sequence of problems you will carry out a proof of Dirichlet's theorem when $m = \ell$, an odd prime. The general outline of the proof is the same as for $m = 8$. Since the characters of $\mathbb{U}_\ell$ are no longer necessarily real-valued (see, for instance, Table 14.1 for the case $\ell = 5$), we will appeal to results in complex analysis rather than real analysis. Having to use complex analysis as opposed to real is more a technical inconvenience than a significant obstacle. We encounter a more serious difficulty when it comes to proving the required nonvanishing of $L(1, \chi)$ for nontrivial χ. For this we follow a clever argument suggested by Sarvadaman Chowla and Louis Mordell.

Characters of $\mathbb{U}_\ell$

14.136 Every character of $\mathbb{U}_\ell$ maps $\mathbb{U}_\ell$ into the group of $(\ell - 1)$th complex roots of unity.

14.137 Recall that $\mathbb{U}_\ell$ is cyclic, and fix a generator g. For every complex $(\ell - 1)$th root of unity ω, there is a unique character χ of $\mathbb{U}_\ell$ with $\chi(g) = \omega$. Thus, there are $\ell - 1$ distinct characters of $\mathbb{U}_\ell$.

14.138 Show that every row of the character table of $\mathbb{U}_\ell$ sums to 0, except the row headed by 1 mod ℓ, which sums to $\ell - 1$. Deduce that for every $a \in \mathbb{Z}$ not divisible by ℓ, the indicator function of $a \bmod \ell \in \mathbb{U}_\ell$ is given by

$$\mathcal{I}_{a \bmod \ell} = \frac{1}{\ell - 1} \sum_\chi \chi(a \bmod \ell)^{-1} \chi.$$

© Springer Nature Switzerland AG 2021
P. Pollack, A. Singha Roy, *Steps into Analytic Number Theory*, Problem Books in Mathematics, https://doi.org/10.1007/978-3-030-65077-3_14

Table 14.1 Character table of $\mathbb{U}_5$

	χ_0	χ_1	χ_2	χ_3
1 mod 5	1	1	1	1
2 mod 5	1	i	-1	$-i$
3 mod 5	1	$-i$	-1	i
4 mod 5	1	-1	1	-1

The L-Functions $L(s, \chi)$

14.139 Let χ be a Dirichlet character modulo ℓ. The same argument you supplied on the last problem set shows that, for all complex s with real part larger than 1, the double series $\sum_p \sum_{k \geq 1} \frac{\chi(p^k)}{kp^{ks}}$ converges absolutely and

$$\exp\left(\sum_p \sum_{k \geq 1} \frac{\chi(p^k)}{kp^{ks}} \right) = L(s, \chi).$$

(The key fact is that, even in the complex world, $\exp(\sum_{k \geq 1} \frac{z^k}{k}) = \frac{1}{1-z}$ whenever $|z| \leq 1, z \neq 1$.) We now *define* (!) "$\mathrm{Log}\, L(s, \chi)$" by the equation

$$\mathrm{Log}\, L(s, \chi) = \sum_{\substack{p \text{ prime} \\ k \geq 1}} \frac{\chi(p^k)}{kp^{ks}}.$$

Verify that the right-hand series in this definition converges uniformly on compact subsets of $\Re(s) > 1$, under any ordering of the terms. Since each summand is analytic, a well-known theorem in complex variables guarantees that $\mathrm{Log}\, L(s, \chi)$ is analytic for $\Re(s) > 1$.

14.140

(a) Show that every column in the character table for $\mathbb{U}_\ell$ sums to 0, except the one headed by χ_0.

(b) Deduce that if χ is a Dirichlet character mod ℓ with $\chi \neq \chi_0$, then $|\sum_{n \leq t} \chi(n)| < \ell$ for all t.

(c) Show that $\sum_{n=1}^{\infty} \frac{\chi(n)}{n^s}$ converges uniformly on compact subsets of $\Re(s) > 0$. Hence, $L(s, \chi)$ is analytic for $\Re(s) > 0$.

14.141 Let χ be a nontrivial Dirichlet character mod ℓ, and suppose that $L(1, \chi) \neq 0$. Since $L(s, \chi)$ is analytic and nonzero in a neighborhood of $s = 1$, a standard theorem in complex variables implies that there is some analytic logarithm of $L(s, \chi)$ in a neighborhood of $s = 1$.

Deduce that our particular logarithm function, $\mathrm{Log}\, L(s, \chi)$ (which is initially defined only when $\Re(s) > 1$), analytically continues to a neighborhood of $s = 1$.

Conclude (assuming $L(1, \chi) \neq 0$): $\mathrm{Log}\, L(s, \chi) = O_\chi(1)$ for $1 < s < 2$.

Reduction to the Nonvanishing of $L(1, \chi)$ for $\chi \neq \chi_0$

It follows from Problem 14.138 that for every integer a not divisible by ℓ, and all $s > 1$,

$$\sum_{p \equiv a \pmod{\ell}} \frac{1}{p^s} = \frac{1}{\ell - 1} \sum_\chi \chi(a)^{-1} \sum_p \frac{\chi(p)}{p^s},$$

where the sums on χ are over all Dirichlet characters $\chi \bmod \ell$.

14.142 For $1 < s < 2$: $\displaystyle\sum_p \frac{\chi_0(p)}{p^s} = \sum_p \frac{1}{p^s} - \sum_{p|\ell} \frac{1}{p^s} = \log \frac{1}{s-1} + O_\ell(1).$

14.143 Estimating $\sum_p \chi(p)p^{-s}$

(a) For each Dirichlet character $\chi \bmod \ell$:

$$\sum_p \frac{\chi(p)}{p^s} = \mathrm{Log}\, L(s, \chi) + O_\chi(1) \quad \text{for } 1 < s < 2.$$

(b) Suppose $\chi \neq \chi_0$ and $L(1, \chi) \neq 0$. Then

$$\sum_p \frac{\chi(p)}{p^s} = O_\chi(1) \quad \text{for } 1 < s < 2.$$

14.144 Assume $L(1, \chi) \neq 0$ for all $\chi \neq \chi_0$. Then for every integer a not divisible by ℓ, and all s with $1 < s < 2$,

$$\sum_{p \equiv a \pmod{\ell}} \frac{1}{p^s} = \frac{1}{\ell - 1} \log \frac{1}{s-1} + O_\ell(1).$$

Consequently, there are infinitely many primes $p \equiv a \pmod{\ell}$.

Nonvanishing of $L(1, \chi)$ for $\chi \neq \chi_0$

14.145 Suppose that $L(1, \chi)$ vanishes for two distinct nontrivial Dirichlet characters $\bmod \ell$, say χ_1 and χ_2.

(a) The quotients $\dfrac{L(s, \chi_1)}{s - 1}$ and $\dfrac{L(s, \chi_2)}{s - 1}$ remain bounded as $s \downarrow 1$, as does

$(s - 1)L(s, \chi_0)$. Deduce: $\dfrac{1}{s - 1} \displaystyle\prod_{\chi} L(s, \chi)$ remains bounded as $s \downarrow 1$.

(b) For $s > 1$: $\displaystyle\sum_{\chi} \mathrm{Log}\, L(s, \chi) = (\ell - 1) \displaystyle\sum_{p^k \equiv 1 \ (\mathrm{mod}\ \ell)} \dfrac{1}{kp^{ks}} \geq 0.$

Hence, $\displaystyle\prod_{\chi} L(s, \chi) \geq 1$, contrary to the conclusion of (a). Thus, there can be

at most one Dirichlet character $\chi \neq \chi_0$ for which $L(1, \chi) = 0$.

14.146 Now consider more closely the situation where there is some (necessarily unique!) nontrivial Dirichlet character χ mod ℓ with $L(1, \chi) = 0$. Show that in this case:

(a) χ assumes only real values,

(b) χ coincides with the Legendre symbol mod ℓ: $\chi(a) = \left(\dfrac{a}{\ell}\right)$ for all a.

Let G be the Gauss sum associated to ℓ (as on Problem 5.43), and let

$$L = L\left(1, \left(\dfrac{\cdot}{\ell}\right)\right) = \sum_{n=1}^{\infty} \left(\dfrac{n}{\ell}\right)\dfrac{1}{n}.$$

We show in the remaining two exercises that $L \neq 0$, completing the proof of Dirichlet's theorem when $m = \ell$.

14.147 Set

$$A = \prod_{0 < a < \ell} \left(1 - e^{2\pi i a/\ell}\right)^{-\left(\frac{a}{\ell}\right)}.$$

Prove: $A = \exp(G \cdot L)$.

14.148 (CHOWLA–MORDELL) Fix a positive integer n with $\left(\frac{n}{\ell}\right) = -1$. Let

$$P(T) = \prod_{0 < r < \ell,\ \left(\frac{r}{\ell}\right)=1} \dfrac{1 - T^{nr}}{1 - T^r} \quad (\in \mathbb{Z}[T]).$$

Assuming $L = 0$, prove:

(a) $e^{2\pi i/\ell}$ is a root of $P(T) - 1$.

(b) $P(T) = 1 + (1 + T + T^2 + \cdots + T^{\ell-1})Q(T)$ for some $Q(T) \in \mathbb{Z}[T]$.

Finally, show that plugging $T = 1$ into the purported identity of (b) leads to a contradiction when considering both sides mod ℓ.

Ingenuity

14.149 Let $\Phi(n)$ denote the number of "Farey fractions" of order n (reduced fractions in $[0, 1]$ with denominator $\leq n$). Then $\{\Phi(n)\}_{n=1}^{\infty}$ begins

$$2, 3, 5, 7, 11, 13, 19, 23, 29, \ldots.$$

Show that despite this promising start, $\Phi(n)$ is infinitely often composite — in fact, infinitely often a multiple of 3.

Note. We do not know if $\Phi(n)$ is prime infinitely often.

Special Step C: Dirichlet's Theorem in the General Case

A careful reading of Special Step B reveals that the primality of $m = \ell$ was used in an essential way only to prove the following.

"Magic" properties of the character table: Each row in the character table for $\mathbb{U}_m$ sums to 0, except for the row headed by 1 mod m, which sums to $\phi(m)$. Each column in the character table for $\mathbb{U}_m$ sums to 0, except for the column headed by χ_0, which sums to $\phi(m)$.

Nonvanishing of $L(1, \chi)$: For all Dirichlet characters χ mod m, with $\chi \neq \chi_0$, we have $L(1, \chi) \neq 0$.

On this final problem set, you will prove that the above assertions hold for arbitrary positive integers m. The remaining arguments of Special Step B then apply to show that for every integer a coprime to m,

$$\sum_{p \equiv a \ (\mathrm{mod}\ m)} \frac{1}{p^s} = \frac{1}{\phi(m)} \log \frac{1}{s-1} + O_m(1).$$

It follows that $\sum_{p \equiv a \ (\mathrm{mod}\ m)} \frac{1}{p}$ diverges (in fact, diverges like $\frac{1}{\phi(m)} \log \log x$— compare with Problem 8.84), and so there are infinitely many primes $p \equiv a$ (mod m).

Character Table Magic

Let G be a finite abelian group.

15.150 (Magic column sums) For each character χ of G, let S_χ be the sum of the character table entries in the column headed by χ. Then $S_{\chi_0} = |G|$. Moreover, for

© Springer Nature Switzerland AG 2021
P. Pollack, A. Singha Roy, *Steps into Analytic Number Theory*, Problem Books in Mathematics, https://doi.org/10.1007/978-3-030-65077-3_15

each character χ and each $g \in G$: $\chi(g)S_\chi = S_\chi$. Consequently: $S_\chi = 0$ if $\chi \neq \chi_0$.

15.151 Suppose that H is a proper subgroup of G, and take $g \in G \setminus H$. Let n be the smallest positive integer with $g^n \in H$.

(a) $H' := H \cup gH \cup g^2 H \cup \cdots \cup g^{n-1} H$ is a subgroup of G with $|H'| = n|H|$.
(b) Every character of H extends in precisely n distinct ways to a character of H'.

15.152 There are exactly $|G|$ characters of G.

15.153 If g is any non-identity element of G, then there is a character χ of G with $\chi(g) \neq 1$.

15.154 (Magic row sums) Let $g \in G$. If S_g is the sum of the entries in the row headed by g, then $S_g = 0$ unless g is the identity, in which case $S_g = |G|$.

Nonvanishing of $L(1, \chi)$ for $\chi \neq \chi_0$

Let $m \in \mathbb{Z}^+$. The same argument employed in Special Step B will show that $L(1, \chi)$ can vanish for at most one nontrivial Dirichlet character χ mod m, which must assume only real values.

Henceforth, χ denotes a real-valued Dirichlet character mod m. Note that since ± 1 are the only real numbers of absolute value 1,

$$\chi(\mathbb{Z}) \subseteq \{-1, 0, 1\}.$$

15.155 For $x \geq 1$: $L(1, \chi) = \sum_{n \leq x} \frac{\chi(n)}{n} + O_m(1/x)$.

15.156 For $x \geq 1$: $\sum_{n \leq x} \frac{\chi(n)}{n} = \int_1^x \left(\sum_{n \leq t} \chi(n) \right) t^{-2} \, dt + O_m(1/x)$.

By Problems 15.155 and 15.156,

$$L(1, \chi) = \int_1^x \left(\sum_{n \leq t} \chi(n) \right) t^{-2} \, dt + O_m(1/x).$$

We now define

$$R(x) = \int_1^x \left(\sum_{n \leq t} \chi(n) \right) \left(\sum_{m \leq x/t} 1 \right) \frac{dt}{t}.$$

15.157 For $x \geq 1$: $R(x) = x \cdot L(1, \chi) + O_m(\log(ex))$.

15.158 $R(x)$ is a weighted sum of the numbers $r(n)$ with $n \leq x$, where

$$r(n) := \sum_{d|n} \chi(d).$$

Precisely:

$$R(x) = \sum_{nm \leq x} \chi(n) \int_n^{x/m} \frac{dt}{t} = \sum_{k \leq x} r(k) \log \frac{x}{k}.$$

15.159 The arithmetic function $r(n)$ from Problem 15.158 is multiplicative. Also: $r(n) \geq 0$ for all $n \in \mathbb{Z}^+$, and $r(n) \geq 1$ if n is a square.

15.160 $R(x) > x^{1/2}$ for all large x.

15.161 $L(1, \chi) \neq 0$.

This proof that $L(1, \chi) \neq 0$ is a modern variant, due to Naoki Yanagisawa, of a classical argument of Franz Mertens.

Ingenuity

15.162 (POWELL, ISRAEL) Let $m, n \in \mathbb{Z}^+$, with $m > 1$. Assume that $(m, n) \neq (2, 1)$. Then $m^p - n$ is composite for infinitely many primes p.

Note. It is an open problem to prove the same assertion when $m = 2$ and $n = 1$.

15.163 Show that the difference

$$\#\{p \leq x : p \equiv 1 \pmod 4\} - \#\{p \leq x : p \equiv 3 \pmod 4\}$$

is not a bounded function of x.

Solutions to Step #1

1.1 We chase through the definitions.

(a) Assume that $f = O(g)$, and choose a constant $C \geq 0$ with $|f| \leq C|g|$. Then $|c \cdot f| = |c| \cdot |f| \leq (|c| \cdot C)|g|$. Hence, $c \cdot f = O(g)$, with one acceptable value of the implied constant being $|c| \cdot C$.

(b) Say $a = O(g)$ and $b = O(h)$. Choose C, D with $|a| \leq C|g|$ and $|b| \leq D|h|$. Then $|ab| = |a||b| \leq CD|g||h| = CD|gh|$. Thus, $ab = O(gh)$.

(c) Say $a = O(f)$ and $b = O(g)$. Choose C, D with $|a| \leq C|f|$ and $|b| \leq D|g|$. Then $|a+b| \leq |a|+|b| \leq \max\{C, D\} \cdot (|f|+|g|)$, and so $a+b = O(|f|+|g|)$.

(d) Since $f = O(g)$, we may choose C with $|f| \leq C|g|$. Now assume that $a = O(f)$ and $b = O(g)$, and choose constants C' and C'' with $|a| \leq C'|f|$ and $|b| \leq C''|g|$. Then

$$|a + b| \leq \max\{C', C''\} \cdot (|f| + |g|) \leq \max\{C', C''\}(C + 1) \cdot |g|.$$

Thus, $a + b = O(g)$.

(e) Choose C, D with $|f| \leq C|g|$ and $|g| \leq D|h|$. Then $|f| \leq CD|h|$, and so $f = O(h)$.

1.2 We first establish the relation $\log(1 + x) = x + O(x^2)$ for the interval $(-0.99, 0.99)$. We know from the theory of Taylor series that when $|x| < 1$,

$$\log(1 + x) = x - \frac{x^2}{2} + \frac{x^3}{3} - \frac{x^4}{4} + \cdots.$$

So for these values of x,

$$\log(1 + x) - x = \left(-\frac{1}{2} + \frac{x}{3} - \frac{x^2}{4} + \cdots\right) x^2.$$

© Springer Nature Switzerland AG 2021
P. Pollack, A. Singha Roy, *Steps into Analytic Number Theory*, Problem Books in Mathematics, https://doi.org/10.1007/978-3-030-65077-3_16

If $|x| < 0.99$, then the parenthesized expression on the right-hand side has absolute value

$$\leq \frac{1}{2} + \frac{|x|}{3} + \frac{|x|^2}{4} + \cdots \leq 1 + |x| + |x|^2 + \cdots = \frac{1}{1 - |x|} < 100.$$

Hence, $\log(1 + x) = x + O(x^2)$ on $(-0.99, 0.99)$, with 100 being an acceptable valued of the implied constant.

Continuing, observe that

$$\lim_{x \to \infty} \frac{\log(1 + x) - x}{x^2} = 0.$$

So for some x_0, and all real numbers $x > x_0$,

$$\left| \frac{\log(1 + x) - x}{x^2} \right| \leq 1, \quad \text{which implies that} \quad |\log(1 + x) - x| \leq |x|^2.$$

Thus, $\log(1 + x) = x + O(x^2)$ on (x_0, ∞), with an implied constant of 1.

Finally, the function $\left| \frac{\log(1+x)-x}{x^2} \right|$ is continuous on the closed and bounded interval $[0.99, x_0]$ and so assumes a maximum value M there. Hence, $\log(1 + x) = x + O(x^2)$ on $[0.99, x_0]$, with M an acceptable implied constant there.

Assembling these estimates shows that $\log(1 + x) = x + O(x^2)$ on $(-0.99, \infty)$, with $\max\{100, M\}$ a possible value of the implied constant. Of course, this shows that $\log(1 + x) = x + O(x^2)$ also on the smaller interval $[0, \infty)$.

To finish off the problem, we *disprove* the relation $\log(1 + x) = x + O(x^2)$ for $(-1, \infty)$. If that relation held, then $\frac{\log(1+x)-x}{x^2}$ would remain bounded as $x \downarrow -1$. But $\lim_{x \downarrow -1} \frac{\log(1+x)-x}{x^2} = -\infty$.

1.3 All of these proofs start the same way. Assume that $f(x) = O(g(x))$ as $x \to \infty$, where $g(x) \to 0$. We can choose a real number x_0 and a nonnegative constant C with $|f(x)| \leq C|g(x)|$ for all $x > x_0$. Since $g(x) \to 0$ as $x \to \infty$, we can also find an $x_1 > x_0$ with $C|g(x)| < \frac{1}{2}$ for $x > x_1$. Then $|f(x)| < \frac{1}{2}$ for $x > x_1$.

We will show that $\frac{1}{1+f(x)} = 1 + O(g(x))$, $e^{f(x)} = 1 + O(g(x))$, and $\log(1 + f(x)) = O(g(x))$ on (x_1, ∞).

Turning to the first claim, we observe that on the interval (x_1, ∞),

$$\frac{1}{1 + f(x)} = 1 - \frac{1}{1 + f(x)} f(x) = 1 + O(f(x)).$$

Here the final equality comes from noting that $\left| \frac{1}{1+f(x)} \right| < 2$ for $x > x_1$. We know that $f(x) = O(g(x))$ for $x > x_1$, and so the $O(f(x))$ term in the last display may be replaced with $O(g(x))$, by Problem 1.1(e).

The other claims are similar. Indeed, for $x > x_1$,

$$e^{f(x)} = 1 + \left(1 + \frac{f(x)}{2!} + \frac{f(x)^2}{3!} + \cdots\right) f(x) = 1 + O(f(x)),$$

since

$$\left|1 + \frac{f(x)}{2!} + \frac{f(x)^2}{3!} + \cdots\right| \le 1 + \frac{1/2}{2!} + \frac{(1/2)^2}{3!} + \cdots = \frac{e^{1/2} - 1}{1/2}.$$

Also,

$$\log(1 + f(x)) = \left(1 - \frac{f(x)}{2} + \frac{f(x)^2}{3} - \frac{f(x)^3}{4} \cdots\right) f(x) = O(f(x)),$$

since

$$\left|1 - \frac{f(x)}{2} + \frac{f(x)^2}{3} - \frac{f(x)^3}{4} + \cdots\right| \le 1 + \frac{1/2}{2} + \frac{(1/2)^2}{3} + \frac{(1/2)^3}{4} + \cdots$$

$$\le 1 + (1/2) + (1/2)^2 + (1/2)^3 + \cdots = 2.$$

Since we can replace $O(f(x))$ with $O(g(x))$, these estimates for $\log(1 + f(x))$ and $e^{f(x)}$ imply the second and third claims.

1.4 We begin by noting that for small t, say $|t| \le \frac{1}{2}$, we have

$$\log(1 + t) = t - \frac{1}{2}t^2 + O(t^3) \tag{1.1}$$

and

$$e^t = 1 + t + O(t^2). \tag{1.2}$$

Both (1.1) and (1.2) are straightforward to prove by manipulations with Taylor series, analogous to those seen in previous solutions. Taking $t = 1/x$ in (1.1), we see that for large x,

$$x \log\left(1 + \frac{1}{x}\right) = x \left(\frac{1}{x} - \frac{1}{2x^2} + O\left(\frac{1}{x^3}\right)\right)$$

$$= 1 - \frac{1}{2x} + O\left(\frac{1}{x^2}\right).$$

Therefore,

$$\left(1 + \frac{1}{x}\right)^x = \exp\left(x \log\left(1 + \frac{1}{x}\right)\right)$$

$$= \exp\left(1 - \frac{1}{2x} + O\left(\frac{1}{x^2}\right)\right)$$

$$= e \cdot \exp\left(-\frac{1}{2x} + O\left(\frac{1}{x^2}\right)\right).$$

Taking t as $-\frac{1}{2x} + O(1/x^2)$ in (1.2) reveals that

$$\exp\left(-\frac{1}{2x} + O\left(\frac{1}{x^2}\right)\right) = 1 - \frac{1}{2x} + O\left(\frac{1}{x^2}\right).$$

(Here the $O(t^2)$ term in (1.2) has been absorbed into the $O(1/x^2)$ error.) Putting this back into the preceding display gives

$$\left(1 + \frac{1}{x}\right)^x = e - \frac{e}{2x} + O\left(\frac{1}{x^2}\right),$$

which was precisely the claim.

Remark. In this problem and the two preceding it, we can avoid bounding tails of series by invoking Lagrange's remainder formula for Maclaurin series approximations: *If f is $(n+1)$-times differentiable on the closed interval from 0 to x, where $x \neq 0$, then*

$$f(x) = f(0) + f'(0)x + \frac{f''(0)}{2!}x^2 + \cdots + \frac{f^{(n)}(0)}{n!}x^n + \frac{f^{(n+1)}(c)}{(n+1)!}x^{n+1}$$

for some c strictly between 0 and x. As one example: For any nonzero $t \in (-1, 1]$, there is some c between 0 and t with $\log(1+t) = t - \frac{1}{2}t^2 + \frac{1}{3}\frac{1}{(1+c)^3}t^3$. This implies (1.1) with $\frac{1}{3}\frac{1}{(1-1/2)^3} = \frac{8}{3}$ as one acceptable value of the implied constant.

1.5 Since $2(f^2 + g^2) - (f + g)^2 = (f - g)^2 \geq 0$, the first statement of the problem is clear. To see that $(f + g)^k = O_k(f^k + g^k)$, introduce the function h whose value at each input is the maximum of the values of f and g. Then

$$(f + g)^k \leq (2h)^k = 2^k h^k \leq 2^k (f^k + g^k).$$

This proves the claimed O-estimate, with an implied constant of 2^k.

1.6 Observe that $a_n = \int_n^{n+1} \frac{1}{n} dx - \int_n^{n+1} \frac{1}{x} dx$ is the area of the region A_n (say) between the graphs of $y = 1/n$ (a horizontal segment) and $y = 1/x$ for $n \leq x \leq n + 1$. Figure 16.1 shows A_1, A_2, A_3, A_4.

We can imagine translating each A_n to the left, to land inside the square $[1, 2] \times [0, 1]$ of area 1. For every positive integer N, the translates of $A_1, \ldots, A_N$ are disjoint, and so

$$1 \geq \sum_{n=1}^{N} \text{area}(A_n) = \sum_{n=1}^{N} a_n.$$

Fig. 16.1 The regions A_1, A_2, A_3, A_4 in the solution to Problem 1.6

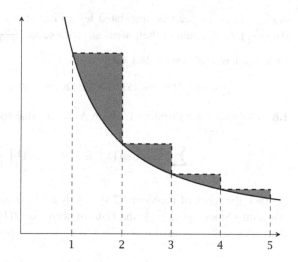

Thus, the partial sums of the infinite series $\sum_{n=1}^{\infty} a_n$ form a bounded, increasing sequence. Therefore, $\sum_{n=1}^{\infty} a_n$ converges.

1.7 By Problem 1.6, it makes sense to define

$$\gamma = \sum_{n=1}^{\infty} a_n.$$

Put $\gamma_N = \sum_{n=1}^{N} a_n$, and notice that

$$\gamma_N = \sum_{n \leq N} \frac{1}{n} - \int_1^{N+1} \frac{dt}{t} = \sum_{n \leq N} \frac{1}{n} - \log(N+1).$$

Rearranging,

$$\sum_{n \leq N} \frac{1}{n} = \log(N+1) + \gamma_N$$

$$= \log(N+1) + \gamma - (\gamma - \gamma_N).$$

The proof will be complete if we show that

$$0 \leq \gamma - \gamma_N \leq \frac{1}{N+1}.$$

The first inequality is clear, as each $a_n \geq 0$. To see the second, notice that $\gamma - \gamma_N = \sum_{n=N+1}^{\infty} a_n = \sum_{n=N+1}^{\infty} \text{area}(A_n)$, in the notation of the last solution. The regions

$A_{N+1}, A_{N+2}, \ldots$ can be translated left, without overlap, into $[N + 1, N + 2] \times [0, \frac{1}{N+1}]$. So the sum of their areas cannot exceed $\frac{1}{N+1}$.

Remark. It is clear from Fig. 16.1 that $\frac{1}{2} < \gamma < 1$. In fact,

$$\gamma = 0.5772156649015328606065120900824024310421593359 3992\ldots$$

1.8 It follows from Problem 1.7 with $N = \lfloor x \rfloor$ that for $x \geq 1$,

$$\sum_{n \leq x} \frac{1}{n} = \log(\lfloor x \rfloor + 1) + \gamma + O\left(\frac{1}{\lfloor x \rfloor + 1}\right).$$

(In fact, the result of Problem 1.7 shows that 1 is an acceptable value of the implied constant.) Since $\frac{1}{\lfloor x \rfloor + 1} \leq \frac{1}{x}$, the O-term above is $O(1/x)$. So it will suffice to show that

$$\log(\lfloor x \rfloor + 1) = \log x + O\left(\frac{1}{x}\right).$$

This follows from

$$0 \leq \log(\lfloor x \rfloor + 1) - \log x = \int_x^{\lfloor x \rfloor + 1} \frac{1}{t} \, dt \leq \frac{1}{x} \int_x^{\lfloor x \rfloor + 1} 1 \, dt \leq \frac{1}{x}.$$

1.9 Square the recurrence relation and rearrange to get that $a_{n+1}^2 - a_n^2 = a_n^{-2} + 2$. Adding from $n = 1$ to N,

$$a_{N+1}^2 = 2N + 1 + \sum_{n=1}^N a_n^{-2}.$$

We immediately deduce that $a_{N+1}^2 \geq 2N + 1$, for each $N \in \mathbb{Z}^+$. This inequality also holds when $N = 0$ (since $a_1 = 1$), and so $a_n^{-2} \leq (2n - 1)^{-1}$ for all $n \in \mathbb{Z}^+$. Referring back to the last displayed equation now gives

$$a_{N+1}^2 \leq 2N + 1 + \sum_{n=1}^N \frac{1}{2n - 1}.$$

Since $\frac{1}{2n-1} \leq \frac{1}{n}$ and $\sum_{n \leq N} \frac{1}{n} = O(\log N)$ for large N,

$$a_{N+1}^2 \leq 2N + O(\log N).$$

Putting this together with the lower bound $a_N^2 \geq 2N - 1$, we deduce that $a_N^2 = 2N + O(\log N)$, as $N \to \infty$.

To finish things off, write

$$2N + O(\log N) = 2N \left(1 + O\left(\frac{\log N}{N}\right)\right).$$

Since $\frac{\sqrt{1+u}-1}{u} \to \frac{1}{2}$ as $u \to 0$, we see that $\sqrt{1+u} = 1 + O(u)$ when $|u|$ is sufficiently small. (This could also be deduced from the Maclaurin series expansion of $(1+u)^{1/2}$.) Thus, as $N \to \infty$,

$$\left(1 + O\left(\frac{\log N}{N}\right)\right)^{1/2} = 1 + O\left(\frac{\log N}{N}\right),$$

and so

$$\sqrt{2N + O(\log N)} = \sqrt{2N} \left(1 + O\left(\frac{\log N}{N}\right)\right) = \sqrt{2N} + O(N^{-1/2} \log N),$$

as desired.

Remark. Problem 1.9 is adapted from [1] (see Problem 60 there).

Reference

1. Donald J. Newman, *A problem seminar*. Problem Books in Mathematics. Springer-Verlag, New York-Berlin, 1982.

Solutions to Step #2

2.10 The function $f(t) = t^{-s}$ is strictly decreasing for $t > 0$, and thus

$$f(n+1) < \int_n^{n+1} f(t)\, dt < f(n)$$

for every natural number n. Since $\int_1^\infty f(t)\, dt$ converges, summing the inequalities from $n = 1$ to ∞ yields

$$\sum_{n=1}^\infty f(n+1) < \int_1^\infty f(t)\, dt = \frac{1}{s-1} < \sum_{n=1}^\infty f(n), \qquad (2.3)$$

In other words,

$$\sum_{n=1}^\infty n^{-s} - 1 < \frac{1}{s-1} < \sum_{n=1}^\infty n^{-s},$$

which immediately implies the desired result.

2.11 We first consider the case when x is a natural number, say $x = m$. With $f(t) = t^{-s}$ as before, we sum the inequality $f(n) < \int_{n-1}^n f(t)\, dt$ over integers $n \geq m + 1$. This yields

$$\sum_{n>m} f(n) = \sum_{n=m+1}^\infty f(n) < \int_m^\infty f(t)\, dt = \frac{m^{1-s}}{s-1}.$$

When $x = m$ is a natural number, this gives the desired result with room to spare (no extra x^{-s} term). If x is not a natural number, apply the displayed inequality with m the least natural number exceeding x. Then

© Springer Nature Switzerland AG 2021
P. Pollack, A. Singha Roy, *Steps into Analytic Number Theory*, Problem Books in
Mathematics, https://doi.org/10.1007/978-3-030-65077-3_17

$$\sum_{n>x} f(n) = f(m) + \sum_{n>m} f(n)$$

$$< m^{-s} + \frac{m^{1-s}}{s-1} < x^{-s} + \frac{x^{1-s}}{s-1},$$

as desired.

2.12 Now let $f(t) = \log t$. This time, f is increasing on $(0, \infty)$, and so $\int_{n-1}^{n} f(t)\, dt \le f(n) \le \int_{n}^{n+1} f(t)\, dt$ for all natural numbers n. So if N is any natural number, then

$$f(1) + f(2) + \cdots + f(N) = f(2) + \cdots + f(N) \ge \int_{1}^{N} \log t\, dt,$$

while

$$f(1) + f(2) + \cdots + f(N) = f(1) + f(2) + \cdots + f(N-1) + \log N$$

$$\le \int_{1}^{N} \log t\, dt + \log N.$$

The last two displayed estimates together imply that

$$\log N! = \log 1 + \cdots + \log N$$

$$= \int_{1}^{N} \log t\, dt + O(\log N).$$

We take $N = \lfloor x \rfloor$ to find that

$$\log \lfloor x \rfloor! = \int_{1}^{x} \log t\, dt - \int_{N}^{x} \log t\, dt + O(\log \lfloor x \rfloor)$$

$$= x \log x - x + \left(1 - \int_{N}^{x} \log t\, dt + O(\log x)\right).$$

Notice that $\int_{N}^{x} \log t\, dt \le (x - N) \log x \le \log x$. Since $\log(ex)$ is larger than both 1 and $\log x$, the entire parenthesized term above is $O(\log(ex))$. This completes the proof.

Why did we write $\log(ex)$ and not $\log x$? One answer: In our proof, we used that $\log(ex) \ge 1$ for all $x \ge 1$, and that inequality is not true without the factor of e. But maybe this is just our proof, and by a more clever argument we could get the result without the e. NO! It is simply not true that for some constant C,

$$|\log \lfloor x \rfloor! - (x \log x - x)| \le C \log x \qquad \text{for all } x \ge 1.$$

You can see this by thinking about values of x slightly larger than 1. For these, the left hand-side will be close to 1, while the right will be close to 0.

Remark. When n is a positive integer, one can obtain much sharper estimates for $\log n!$. For example, an explicit form of "Stirling's approximation" states that for all $n \in \mathbb{Z}^+$,

$$\sqrt{2\pi n}(n/e)^n e^{\frac{1}{12n+1}} < n! < \sqrt{2\pi n}(n/e)^n e^{\frac{1}{12n}}.$$

This formula determines $\log n!$ to within $O(1/n^2)$, whereas the result of Problem 2.12 only determines $\log n!$ to within $O(\log n)$. Nevertheless, there is a sense in which the estimate of Problem 2.12 is optimal. Namely, when viewing $\log \lfloor x \rfloor!$ as a function of a real number x, there are jumps of size $\log x$ at all positive integers x. So one cannot hope for a continuous approximation to $\log \lfloor x \rfloor!$ with an error of size smaller than $O(\log x)$.

2.13 If p is any of $p_1, \ldots, p_k$, then p divides exactly one of a or b, and so does not divide $a+b$. But $a+b$ is an integer larger than 1, and so has some prime factor. Thus, we have located a prime not among $p_1, \ldots, p_k$. Since this holds for any conceivable finite list of primes $p_1, \ldots, p_k$, there must be infinitely many primes.

2.14 Let $F_n := 2^{2^n} + 1$ denote the nth Fermat number. We begin by noting that

$$F_{n+1} - 2 = \prod_{i=0}^{n} F_i$$

for all nonnegative integers n. Indeed, the identity is easily checked for $n = 0$, and if it holds for some nonnegative integer n, then $F_{n+2} - 2 = 2^{2^{n+2}} - 1 = (2^{2^{n+1}} - 1)(2^{2^{n+1}} + 1) = (\prod_{i=0}^{n} F_i)F_{n+1} = \prod_{i=0}^{n+1} F_i$. So the result follows by induction.

An immediate consequence is that $\gcd(F_i, F_j) = 1$ whenever $i \neq j$: For if (say) $i > j$, then $F_i - 2 \equiv 0 \pmod{F_j}$. Thus, $\gcd(F_i, F_j) = \gcd((F_i - 2) + 2, F_j) = \gcd(2, F_j) = 1$. In the last step, we used that $F_j = 2^{2^j} + 1$ is odd.

Finally as each of the Fermat numbers is greater than one and no two distinct Fermat numbers share a common prime divisor, the infinitude of the primes follows from that of the Fermat numbers.

2.15 An integer in $[1, N]$ which is not squarefree is divisible by p^2 for some prime p. For each p, the number of integers in $[1, N]$ divisible by p^2 is at most $\frac{N}{p^2}$. Thus, the number of non-squarefree integers in $[1, N]$ is at most $\sum_p \frac{N}{p^2}$, where the sum on p extends over all primes. Consequently, the number of integers in $[1, N]$ that *are* squarefree is at least

$$N - \sum_p \frac{N}{p^2} \geq N - \sum_{m \geq 2} \frac{N}{m^2} = cN$$

where

$$c = 1 - \sum_{m \geq 2} \frac{1}{m^2} > 1 - \sum_{m \geq 2} \frac{1}{m(m-1)} = 1 - \sum_{m \geq 2} \left(\frac{1}{m-1} - \frac{1}{m} \right) = 0.$$

The infinitude of primes is again immediate: If there are only k different primes $p_1, \ldots, p_k$, then there are only 2^k squarefree numbers, namely the products $\prod_{i \in I} p_i$ for subsets $I \subseteq \{1, 2, \ldots, k\}$. But there are infinitely many squarefree numbers, as we see by sending N to infinity in the above estimate.

2.16 The natural numbers $n \leq x$ divisible only by primes in $\mathcal{P}$ are in one-to-one correspondence with the tuples $(e_1, \ldots, e_k)$ of nonnegative integers satisfying $\prod_{i=1}^{k} p_i^{e_i} \leq x$. Upon taking logarithms, this last inequality becomes

$$\sum_{i=1}^{k} e_i \log p_i \leq \log x.$$

Thus, our counting problem is equivalent to the following exercise in discrete geometry: Estimate the number of lattice points $(e_1, \ldots, e_k)$ belonging to the simplex

$$S = \{(x_1, \ldots, x_k) \in \mathbb{R}^k : \sum_{i=1}^{k} x_i \log p_i \leq \log x, \text{ and all } x_i \geq 0\}.$$

It is helpful to view $\mathbb{R}^k$ as partitioned into unit hypercubes having their vertices at lattice points. Explicitly, for each $(e_1, \ldots, e_k) \in \mathbb{Z}^k$, we let $C_{(e_1, \ldots, e_k)}$ denote the unit hypercube whose lower 'lower left' corner is $(e_1, \ldots, e_k)$. That is,

$$C_{(e_1, \ldots, e_k)} = \{(x_1, \ldots, x_k) \in \mathbb{R}^k : \text{each } e_i \leq x_i < e_i + 1\}.$$

Then every $(x_1, \ldots, x_k) \in S$ belongs to a cube $C_{(e_1, \ldots, e_k)}$ for some lattice point $(e_1, \ldots, e_k) \in S$. In the other direction, if $(e_1, \ldots, e_k)$ is a lattice point belonging to S, then every $(x_1, \ldots, x_k) \in C_{(e_1, \ldots, e_k)}$ satisfies

$$\sum_{i=1}^{k} x_i \log p_i \leq \sum_{i=1}^{k} (e_i + 1) \log p_i \leq \log(x p_1 \cdots p_k),$$

so that $(x_1, \ldots, x_k)$ belongs to the simplex

$$S' = \{(x_1, \ldots, x_k) \in \mathbb{R}^k : \sum_{i=1}^{k} x_i \log p_i \leq \log(x p_1 \cdots p_k), \text{ and all } x_i \geq 0\}.$$

Hence,

$$S \subseteq \bigcup_{(e_1,\ldots,e_k) \in \mathbb{Z}^K \cap S} C_{(e_1,\ldots,e_k)} \subseteq S'.$$

Comparing (hyper)volumes, we see that if N is our lattice point count, then

$$\mathrm{vol}(S) \leq N \leq \mathrm{vol}(S').$$

It is a standard exercise in multivariable calculus to compute the (hyper)volumes of multidimensional tetrahedra:

$$\mathrm{vol}(S) = \frac{(\log x)^k}{k! \prod_{i=1}^{k} \log p_i},$$

while

$$\mathrm{vol}(S') = \frac{(\log x + \log(p_1 \cdots p_k))^k}{k! \prod_{i=1}^{k} \log p_i}$$

$$= \frac{(\log x)^k}{k! \prod_{i=1}^{k} \log p_i} + \sum_{j=1}^{k} \frac{\binom{k}{j} (\log x)^{k-j} \log(p_1 \cdots p_k)^j}{k! \prod_{i=1}^{k} \log p_i}.$$

For each $j = 1, 2, \ldots, k - 1$, we have $(\log x)^{k-j} \leq \log(ex)^{k-j} \leq \log(ex)^{k-1}$. Thus, each term in the sum on j is $O(\log(ex)^{k-1})$, where the implied constant may depend on k and the p_i. Hence, that entire sum is $O_P(\log(ex)^{k-1})$, and we have the estimate claimed in the problem statement.

The infinitude of primes is again an easy consequence. Indeed, if $p_1, \cdots, p_k$ are all the primes, then no number has a prime factor outside $\{p_1, \ldots, p_k\}$. Hence, the total count of natural numbers up to x is $O((\log x)^{k-1})$, as $x \to \infty$. But that count is $\lfloor x \rfloor$, and $\lfloor x \rfloor$ is not $O((\log x)^{k-1})$. Contradiction.

2.17 We fix $n \in \mathbb{Z}^+$ and proceed by induction on r. The result is immediate for $r = 0$. Suppose that for some r with $0 \leq r < n$ we have $\sum_{j=0}^{r} (-1)^j \binom{n}{j} = (-1)^r \binom{n-1}{r}$. Then

$$\sum_{j=0}^{r+1} (-1)^j \binom{n}{j} = \sum_{j=0}^{r} (-1)^j \binom{n}{j} + (-1)^{r+1} \binom{n}{r+1}$$

$$= (-1)^r \binom{n-1}{r} + (-1)^{r+1} \binom{n}{r+1}$$

$$= (-1)^r \left(\binom{n-1}{r} - \binom{n}{r+1} \right)$$

$$= (-1)^{r+1} \binom{n-1}{r+1}.$$

Here the last step is a consequence of Pascal's identity in the form $\binom{n}{r+1} = \binom{n-1}{r} + \binom{n-1}{r+1}$.

Alternative Solution An application of generating functions yields a one-line solution. We observe that on the one hand, the right hand side $(-1)^r \binom{n-1}{r}$ is the coefficient of x^r in $(1-x)^{n-1}$, whereas on the other,

$$(1-x)^{n-1} = (1-x)^n (1-x)^{-1} = \left(\sum_{i=0}^{n} (-1)^i \binom{n}{i} x^i \right) \left(\sum_{j \geq 0} x^j \right)$$

and the coefficient of x^r in the rightmost expression is equal to

$$\sum_{\substack{i+j=r, \\ i,j \geq 0}} (-1)^i \binom{n}{i} = \binom{n}{0} - \binom{n}{1} + \cdots + (-1)^r \binom{n}{r}.$$

2.18 The inclusion-exclusion formula (or "Principle of Inclusion-Exclusion") states that

$$|A \setminus (A_1 \cup \cdots \cup A_k)| = |A| - \sum_{1 \leq i_1 \leq k} |A_{i_1}| + \sum_{1 \leq i_1 < i_2 \leq k} |A_{i_1} \cap A_{i_2}|$$

$$- \sum_{1 \leq i_1 < i_2 < i_3 \leq k} |A_{i_1} \cap A_{i_2} \cap A_{i_3}| + \cdots + (-1)^k |A_1 \cap \cdots \cap A_k|. \qquad (2.4)$$

For the proof, we show that the two sides of (2.4) count each $x \in A$ the same number of times.

Case 1: $x \notin A_1 \cup \cdots \cup A_k$. Then x is counted once in $|A \setminus (A_1 \cup \cdots \cup A_k)|$, once in $|A|$, and nowhere else among the other cardinalities appearing in (2.4). Thus, x is counted exactly once on each side of (2.4).

Case 2: $x \in A_1 \cup \cdots \cup A_k$. In this case, x is not counted on the left hand side of (2.4). Let n be the number of indices i, $1 \leq i \leq k$, for which $x \in A_i$, so that $n \geq 1$. For each $1 \leq r \leq n$, the element x is counted exactly $\binom{n}{r}$ times in the sum $\sum_{1 \leq i_1 < \cdots < i_r \leq k} |A_{i_1} \cap \cdots \cap A_{i_r}|$. Thus, the number of times x is counted by the right hand side of (2.4) is precisely

$$1 - \binom{n}{1} + \binom{n}{2} + \cdots + (-1)^n \binom{n}{n} = \sum_{j=0}^{n} (-1)^j \binom{n}{j} = (1-1)^n = 0,$$

establishing the desired result in this case.

The name "inclusion-exclusion" is explained by the form of the terms on the right of (2.4). The first term "includes" all elements of A, the second "excludes" those belonging to each A_i, the third "includes" those belonging to each pair $A_i \cap A_j$ with $i < j$, etc.

2.19 The form of the identity suggests an application of the Principle of Inclusion-Exclusion. Let $p_1 < p_2 < \cdots < p_k$ be the primes not exceeding $\sqrt{x}$. Let A be the set of natural numbers not exceeding x, and let A_i be the subset of A consisting of the multiples of p_i. Then

$$A \setminus (A_1 \cup \cdots \cup A_k) = \{n \le x : n \text{ is not divisible by any prime } p \le \sqrt{x}\}$$
$$= \{1\} \cup \{p \text{ prime} : \sqrt{x} < p \le x\},$$

so that

$$|A \setminus (A_1 \cup \cdots \cup A_k)| = \pi(x) - \pi(\sqrt{x}) + 1.$$

The same cardinality can be found with (2.4). Clearly, $|A| = \lfloor x \rfloor$. If $1 \le i_1 < \cdots < i_r \le k$, then $|A_{i_1} \cap \cdots \cap A_{i_r}|$ counts the number of multiples of $p_{i_1} \cdots p_{i_r}$ in $[1, x]$, which is $\left\lfloor \frac{x}{p_{i_1} \cdots p_{i_r}} \right\rfloor$. Inserting these expressions into (2.4) yields the claimed result.

2.20 Suppose to the contrary that f is nonconstant. Let $p = f(1)$; by hypothesis, p is prime. Then for each natural number t, we have that

$$f(1 + pt) \equiv f(1) \equiv 0 \pmod{p}.$$

But $f(1 + pt)$ is also prime, and so $f(1 + pt) = p$. This holds for every t, and so $f(1 + pT) - p$ is a nonzero polynomial in T with infinitely many roots, an absurdity!

2.21 The result is obvious when k is even, so we suppose that k is odd. Write

$$k - 1 = 2^t q$$

for natural numbers t, q with q odd. (We use here that $k > 1$!) Fix an integer $n \ge t$. Then $2^{t+1} \mid 2^{2^n}$, but $2^{t+1} \nmid k - 1$. Hence, $2^{2^n} + k \not\equiv 1 \pmod{2^{t+1}}$ and so there must be a prime p dividing $2^{2^n} + k$ with $p \not\equiv 1 \pmod{2^{t+1}}$. It suffices to show that there are infinitely many positive integers m for which

$$p \mid 2^{2^m} + k.$$

To prove this, we start by observing that $2^{2^n} \equiv -k \pmod{p}$. Thus, for each integer $m \ge n$,

$$2^{2^m} + k \equiv 0 \pmod{p} \iff 2^{2^m} \equiv 2^{2^n} \pmod{p}$$
$$\iff 2^{2^m - 2^n} \equiv 1 \pmod{p}$$
$$\iff \ell \mid 2^n (2^{m-n} - 1),$$

where ℓ denotes the multiplicative order of 2 modulo p. Write $\ell = 2^{t'}q'$, with q' odd, and notice that $2^{t'} \mid \ell \mid p - 1$, so that $t' \le t \le n$. Hence,

$$\ell = 2^{t'}q' \mid 2^n(2^{m-n} - 1)$$

whenever q' divides $2^{m-n} - 1$. And this happens for infinitely many integers m, for example all $m = n + u\phi(q')$ for some positive integer u.

Remark. It can be surprisingly difficult to prove that naturally defined sequences do *not* contain only primes (or only primes from some point on).

A famous example is provided by the sequence of "Fermat numbers" $2^{2^n} + 1$. The first several terms of this sequence are prime (starting at $n = 0$): 3, 5, 17, 257, 65537. Fermat's correspondence records his conjecture—expressed multiple times, with varying degrees of certitude—that this pattern continues forever. But around 1732, Euler discovered that

$$2^{2^5} + 1 = 641 \cdot 6700417.$$

Time has not been kind to Fermat's conjecture. We know today that $2^{2^n} + 1$ is composite for each n with $5 \le n \le 32$. It may well be that $2^{2^n} + 1$ is composite for all $n \ge 5$. But we still do not know how to *dis*prove the asymptotic version of Fermat's conjecture, that $2^{2^n} + 1$ is prime for all sufficiently large n.

In this connection, we mention an amusing elementary result of Schinzel [3]: Either there are infinitely many composite numbers of the form $2^{2^n} + 1$, or there are infinitely many composite numbers of the form $6^{2^n} + 1$. The reader is encouraged to look for their own proof of this result before consulting [3].

Here is another family of stubborn sequences. Let r be a rational number, $r > 1$. It should certainly be the case that $\lfloor r^n \rfloor$ contains infinitely many composite terms. This has been proved in (finitely many) certain special cases, including $r = 4/3$ and $r = 7/5$, in [2] and [1]. Can it be proved in general?

References

1. Artūras Dubickas and Aivaras Novikas, *Integer parts of powers of rational numbers*. Math. Z. 251 (2005), 635–648.
2. William Forman and Harold N. Shapiro, *An arithmetic property of certain rational powers*. Comm. Pure Appl. Math. 20 (1967), 561–573.
3. Andrzej Schinzel, *Remarque au travail de W. Sierpiński sur les nombres $a^{2^n} + 1$*. Colloq. Math. 10 (1963), 137–138.

Solutions to Step #3

3.22 The integral test shows the convergence of $\sum_{n=1}^{\infty} \frac{1}{n^s}$, for every $s > 1$. As a series of positive terms, this convergence is necessarily absolute; thus, the terms can be grouped or rearranged without affecting the value of the sum.

To get started, we use that each positive integer n has a unique representation as $n = 2^{e_1} n_1$, where $e_1 \geq 0$ and n_1 is odd. Thus,

$$\sum_{n=1}^{\infty} \frac{1}{n^s} = \sum_{e_1 \geq 0} \sum_{\substack{n_1 \geq 1 \\ \gcd(n_1,2)=1}} \frac{1}{(2^{e_1} n_1)^s} = \sum_{e_1 \geq 0} \frac{1}{2^{e_1 s}} \sum_{\substack{n_1 \geq 1 \\ \gcd(n_1,2)=1}} \frac{1}{n_1{}^s}$$

$$= \left(\sum_{\substack{n_1 \geq 1 \\ \gcd(n_1,2)=1}} \frac{1}{n_1{}^s} \right) \left(1 + \frac{1}{2^s} + \frac{1}{2^{2s}} + \cdots \right).$$

Now we iterate. Each odd n_1 can be written uniquely as $3^{e_2} n_2$, where n_2 is coprime with 6. Hence,

$$\sum_{n=1}^{\infty} \frac{1}{n^s} = \left(\sum_{\substack{n_2 \geq 1 \\ \gcd(n_2,6)=1}} \frac{1}{n_2{}^s} \right) \left(1 + \frac{1}{2^s} + \frac{1}{2^{2s}} + \cdots \right) \left(1 + \frac{1}{3^s} + \frac{1}{3^{2s}} + \cdots \right).$$

Continuing, we find that for an arbitrary integer $N \geq 2$,

$$\sum_{n=1}^{\infty} \frac{1}{n^s} = \left(\sum_{\substack{m \geq 1 \\ \gcd(m, \prod_{p \leq N} p)=1}} \frac{1}{m^s} \right) \prod_{p \leq N} \left(1 + \frac{1}{p^s} + \frac{1}{p^{2s}} + \cdots \right).$$

We send $N \to \infty$. The sum on m tends to 1, since

© Springer Nature Switzerland AG 2021
P. Pollack, A. Singha Roy, *Steps into Analytic Number Theory*, Problem Books in
Mathematics, https://doi.org/10.1007/978-3-030-65077-3_18

$$1 \leq \sum_{\substack{m \geq 1 \\ \gcd(m, \prod_{p \leq N} p)=1}} \frac{1}{m^s} \leq 1 + \sum_{m > N} \frac{1}{m^s}$$

and $\sum_{m > N} \frac{1}{m^s}$ is the tail of a convergent series. We conclude that

$$\sum_{n=1}^{\infty} \frac{1}{n^s} = \prod_{p} \left(1 + \frac{1}{p^s} + \frac{1}{p^{2s}} + \cdots \right),$$

as desired. The final equality in the problem statement follows from the geometric series identity $1 + \frac{1}{p^s} + \frac{1}{p^{2s}} + \cdots = \frac{1}{1 - \frac{1}{p^s}}$.

3.23 The previous problem allows us to write $\log \zeta(s) = \sum_p \log \frac{1}{1 - \frac{1}{p^s}}$. Recalling that

$$\log \frac{1}{1 - x} = -\log(1 - x) = x + \frac{1}{2}x^2 + \frac{1}{3}x^3 + \cdots$$

whenever $|x| < 1$, we find that

$$\log \zeta(s) = \sum_p \sum_{k \geq 1} \frac{1}{k p^{ks}},$$

which is the first equality claimed in the problem statement. To obtain the second, it is enough to show that

$$\sum_p \sum_{k \geq 2} \frac{1}{k p^{ks}} = O(1).$$

In fact (keeping in mind that $s > 1$),

$$\sum_p \sum_{k \geq 2} \frac{1}{k p^{ks}} < \sum_p \sum_{k \geq 2} \frac{1}{p^k} = \sum_p \frac{1}{p(p-1)} < \sum_{m \geq 2} \frac{1}{m(m-1)} = 1.$$

3.24 By Problem 2.10, if $1 < s < 2$ then

$$\log \frac{1}{s-1} < \log \zeta(s) < \log \frac{s}{s-1} < \log 2 + \log \frac{1}{s-1},$$

so that

$$\log \zeta(s) = \log \frac{1}{s-1} + O(1).$$

By Problem 3.23,

$$\log \zeta(s) = \sum_p \frac{1}{p^s} + O(1).$$

Comparing the estimates,

$$\sum_p \frac{1}{p^s} = \log \frac{1}{s-1} + O(1).$$

As $s \downarrow 1$, we see that $\log \frac{1}{s-1} \to \infty$ and so $\sum_p \frac{1}{p^s} \to \infty$. This certainly implies that $\sum_p \frac{1}{p}$ diverges: If that series converged to A (say), we would have that $\sum_p \frac{1}{p^s} \le A$ for every $s > 1$.

3.25 By Problem 3.22, the identity $\zeta(s)^{-1} = \prod_p (1 - \frac{1}{p^s})$ holds for every $s > 1$. So it will be enough to find a sequence $\{c(n)\}$ satisfying

$$\prod_p \left(1 - \frac{1}{p^s}\right) = \sum_{n=1}^{\infty} \frac{c(n)}{n^s} \tag{3.5}$$

for all $s > 1$. If we formally expand out the left-hand side ("formally" meaning that we ignore issues of convergence), we obtain a weighted sum of terms $\frac{1}{n^s}$, where the (only) positive integers n that appear are the products of distinct primes, and where $\frac{1}{n^s}$ appears with a coefficient of $+1$ or -1 according to whether n is the product of an even or an odd number of primes. This suggests (but does not prove) that we can take

$$c(n) = \begin{cases} 0 & \text{unless } n \text{ is squarefree,} \\ (-1)^k & \text{if } n = p_1 \cdots p_k \text{ with the } p_i \text{ distinct.} \end{cases} \tag{3.6}$$

Note that $c(1) = 1$ with this definition, as 1 has a prime factorization into $k = 0$ distinct primes! Moreover, $c(n)$ is multiplicative with the usual definition from elementary number theory: $c(ab) = c(a)c(b)$ whenever $\gcd(a, b) = 1$.

To rigorously prove that (3.5) holds with $c(n)$ defined as above, we imitate the solution to Problem 3.22. The series on the right of (3.5) converges absolutely when $s > 1$, and so we may rearrange or regroup terms at will. Writing $n = 2^{e_1} n_1$, where $e_1 \ge 0$ and n_1 is odd,

$$\sum_{n=1}^{\infty} \frac{c(n)}{n^s} = \sum_{e_1 \ge 0} \frac{c(2^{e_1})}{2^{e_1 s}} \sum_{\substack{n_1 \ge 1 \\ \gcd(n_1, 2) = 1}} \frac{c(n_1)}{n_1^s}$$

$$= \left(\sum_{\substack{n_1 \geq 1 \\ \gcd(n_1, 2) = 1}} \frac{c(n_1)}{n_1{}^s} \right) \left(1 + \frac{c(2)}{2^s} + \frac{c(4)}{4^s} + \cdots \right)$$

$$= \left(\sum_{\substack{n_1 \geq 1 \\ \gcd(n_1, 2) = 1}} \frac{c(n_1)}{n_1{}^s} \right) \left(1 - \frac{1}{2^s} \right).$$

Iterating this procedure as in Problem 3.22, we find that for every $N \geq 2$,

$$\sum_{n=1}^{\infty} \frac{c(n)}{n^s} = \left(\sum_{\substack{m \geq 1 \\ \gcd(m, \prod_{p \leq N} p) = 1}} \frac{c(m)}{m^s} \right) \prod_{p \leq N} \left(1 - \frac{1}{p^s} \right). \tag{3.7}$$

The sum on m tends to 1 as $N \to \infty$, since

$$\left| \left(\sum_{\substack{m \geq 1 \\ \gcd(m, \prod_{p \leq N} p) = 1}} \frac{c(m)}{m^s} \right) - 1 \right| \leq \sum_{m > N} \frac{|c(m)|}{m^s},$$

with the right-hand sum being the tail of a convergent series. Finally, taking $N \to \infty$ in (3.7) yields (3.5).

The reader will have noticed that $c(n)$ is precisely the "Möbius function", familiar from elementary number theory. In future problem sets, it will be denoted $\mu(n)$ as usual.

Alternative Solution Suppose that $\sum_{n=1}^{\infty} \frac{c(n)}{n^s}$ converges absolutely for all real $s > 1$. Then, for these same values of s,

$$\zeta(s) \sum_{n=1}^{\infty} \frac{c(n)}{n^s} = \sum_{m=1}^{\infty} \frac{1}{m^s} \sum_{n=1}^{\infty} \frac{c(n)}{n^s} = \sum_{m,n \geq 1} \frac{c(n)}{(mn)^s}$$

$$= \sum_{N=1}^{\infty} \sum_{mn=N} \frac{c(n)}{N^s} = \sum_{N=1}^{\infty} \left(\sum_{n \mid N} c(n) \right) \frac{1}{N^s}$$

$$= g(1) + \frac{g(2)}{2^s} + \frac{g(3)}{3^s} + \cdots,$$

where

$$g(N) := \sum_{n \mid N} c(n).$$

This leads us to ask: Under what conditions on the sequence $g(1), g(2), g(3), \ldots$ is

$$g(1) + \frac{g(2)}{2^s} + \frac{g(3)}{3^s} + \cdots = 1$$

for all $s > 1$? Certainly this holds if $g(1) = 1$ and $g(N) = 0$ for $N > 1$. (Later we will prove the less obvious fact that this is the only way that equality can hold for all $s > 1$.)

Thus, we will be done if $\{c(n)\}$ can be chosen to satisfy

$$\sum_{n|N} c(n) = \begin{cases} 1 & \text{if } N = 1, \\ 0 & \text{otherwise,} \end{cases} \tag{3.8}$$

and also to make the series $\sum_{n=1}^{\infty} \frac{c(n)}{n^s}$ converge absolutely whenever $s > 1$. Is this possible? Yes! It is well known from elementary number theory that the Möbius function, defined by (3.6), satisfies (3.8). Since $|c(n)| \leq 1$ for all n, the absolute convergence condition also holds. So $c(n)$ as defined by (3.6) solves our problem.

Remark. The method used in the first solution, and in Problem 3.22, is easily adapted to prove the following very useful general theorem concerning series of the form $\sum_{n=1}^{\infty} \frac{f(n)}{n^s}$: Let f be *a complex-valued multiplicative function with* $f(1) = 1$. *Suppose that* $\sum_{n=1}^{\infty} \frac{f(n)}{n^s}$ *converges absolutely, for a certain* $s \in \mathbb{C}$. *For this same* s,

$$\sum_{n=1}^{\infty} \frac{f(n)}{n^s} = \prod_p \left(1 + \frac{f(p)}{p^s} + \frac{f(p^2)}{p^{2s}} + \cdots \right).$$

3.26 We imitate the solution to Problem 2.18.

Why does stopping after an inclusion always give an overcount? It suffices to show that when $r \geq 0$ is even, each $x \in A$ is counted at least as many times by

$$|A| - \sum_{1 \leq i_1 \leq k} |A_{i_1}| + \sum_{1 \leq i_1 < i_2 \leq k} |A_{i_1} \cap A_{i_2}|$$

$$+ \cdots + (-1)^r \sum_{1 \leq i_1 < \cdots < i_r \leq k} |A_{i_1} \cap \cdots \cap A_{i_r}| \tag{3.9}$$

as it is by $|A \setminus (A_1 \cup \cdots \cup A_k)|$. If x is not in any A_i, then x is counted once by both expressions. If instead x belongs to A_i for $n \geq 1$ values of i, then x is not counted in $|A \setminus (A_1 \cup \cdots \cup A_k)|$, while the number of times x is counted by (3.9) is

$$\sum_{j=0}^{r} (-1)^j \binom{n}{j} = (-1)^r \binom{n-1}{r} \geq 0.$$

Here we use that r is even to obtain the final inequality.

The proof that we undercount when stopping after an exclusion is the same; this time r is odd, and so the final inequality in the last display is reversed.

3.27 Suppose that $\mathcal{A}$ does not contain 0% of the positive integers. Then for some $\delta > 0$, there are arbitrarily large values of x with

$$\sum_{n \leq x,\ n \in \mathcal{A}} 1 > \delta \sum_{n \leq x} 1.$$

Whenever this inequality holds, it also holds with x replaced by its integer part. So there are arbitrarily large $N \in \mathbb{Z}^+$ for which

$$\sum_{n \leq N,\ n \in \mathcal{A}} 1 > \delta N.$$

Letting $\eta = \frac{1}{2}\delta$, we see that for such N,

$$\sum_{\substack{a \in \mathcal{A} \\ \eta N < a \leq N}} \frac{1}{a} \geq \frac{1}{N} \sum_{\substack{a \in \mathcal{A} \\ \eta N < a \leq N}} 1 = \frac{1}{N} \left(\sum_{\substack{a \in \mathcal{A} \\ a \leq N}} 1 - \sum_{\substack{a \in \mathcal{A} \\ a \leq \eta N}} 1 \right)$$

$$> \frac{1}{N} (\delta N - \eta N) = \frac{1}{2}\delta. \tag{3.10}$$

This contradicts that $\sum_{a \in \mathcal{A}} \frac{1}{a}$ converges. Indeed, if that sum converges to L (say), then as $N \to \infty$,

$$\sum_{\substack{a \in \mathcal{A} \\ \eta N < a \leq N}} \frac{1}{a} = \sum_{\substack{a \in \mathcal{A} \\ a \leq N}} \frac{1}{a} - \sum_{\substack{a \in \mathcal{A} \\ a \leq \eta N}} \frac{1}{a} \to L - L = 0,$$

so that (3.10) cannot hold for arbitrarily large N.

Alternative Solution What we are asked is to do is equivalent to proving that $\frac{1}{x} \sum_{a \in \mathcal{A},\ a \leq x} 1 \to 0$, as $x \to \infty$. The key is to observe that $1 \leq x/a$ for every $a \leq x$. Hence,

$$\sum_{\substack{a \in \mathcal{A} \\ a \leq x}} 1 \leq \sqrt{x} + \sum_{\substack{a \in \mathcal{A} \\ \sqrt{x} < a \leq x}} 1 \leq \sqrt{x} + x \sum_{\substack{a \in \mathcal{A} \\ a > \sqrt{x}}} \frac{1}{a}.$$

In the last step we replaced 1 with x/a and we extended the sum over all $a > \sqrt{x}$, which only serves to make the sum larger. Now we notice that $\sum_{a \in \mathcal{A},\ a > \sqrt{x}} 1/a \to 0$, as the tail of a convergent series. So the desired result follows from the last display, upon dividing by x and sending $x \to \infty$.

3.28 The integral test (or Cauchy condensation) implies that $\sum_m \frac{1}{m(\log m)^{1.01}}$ converges. So if we were to have $a_m \geq m(\log m)^{1.01}$ for all large m, then $\sum_m \frac{1}{a_m}$ would converge by the comparison test. So there must be infinitely many m with $a_m < m(\log m)^{1.01}$.

Write $A(t)$ for $\sum_{n \leq t,\ n \in A} 1$. Whenever $a_m < m(\log m)^{1.01}$,

$$m = A(a_m) \leq A(m(\log m)^{1.01}),$$

so that if m is a large number of this kind and $x := m(\log m)^{1.01}$,

$$A(x) \geq m = x/(\log m)^{1.01} > x/(\log x)^{1.01},$$

as desired.

Many other functions can take the place of $x/(\log x)^{1.01}$. For instance, running the above argument with $m(\log m)^{1.01}$ replaced by $m(\log m)(\log \log m)^{1.01}$ or $m(\log m)(\log \log m)(\log \log \log m)^{1.01}$ allows us to take $x/(\log x (\log \log x)^{1.01})$ or $x/(\log x (\log \log x)(\log \log \log x)^{1.01})$ here.

3.29 The key is to recognize that $\sum_{d|n} g(d)h(n/d) = \sum_{de=n} g(d)h(e)$. Thus,

$$\sum_{n \leq x} \sum_{d|n} g(d)h(n/d) = \sum_{n \leq x} \sum_{de=n} g(d)h(e) = \sum_{de \leq x} g(d)h(e).$$

Here the final sum is over all ordered pairs (d, e) of positive integers with $de \leq x$. We can interpret this sum on pairs as a double sum by choosing to first sum on d or on e. In the former case, the e corresponding to a given d are precisely those with $e \leq x/d$, so that

$$\sum_{de \leq x} g(d)h(e) = \sum_{d \leq x} g(d) \sum_{e \leq x/d} h(e).$$

This yields the first equality claimed in the problem. The second is obtained similarly by first summing on e.

Remark. We can recast the above solution in geometric terms. A sum on $de \leq x$ can be viewed as a sum over first-quadrant lattice points in the (d, e)-plane that lie on or below the hyperbola $de = x$. Summing on d (respectively, e) first corresponds to grouping together points on vertical (respectively, horizontal) lines.

3.30 Taking g and h to be identically 1, the divisor function $\tau(n)$ satisfies $\sum_{d|n} g(d)h(n/d) = \tau(n)$ for all $n \in \mathbb{Z}^+$. So by the previous problem,

$$\sum_{n\le x} \tau(n) = \sum_{ab\le x} 1 = \sum_{a\le x} 1 \sum_{b\le x/a} 1 = \sum_{a\le x} \left\lfloor \frac{x}{a} \right\rfloor$$

$$= \sum_{a\le x} \left(\frac{x}{a} + O(1)\right) = x \sum_{a\le x} \frac{1}{a} + O(x) = x \log x + O(x).$$

Here the last equality is a consequence of Problem 1.8.

Remark. A more precise estimate for $\sum_{n\le x} \tau(n)$ was obtained by Dirichlet in 1849.

Dirichlet's starting point is the observation that $ab \le x$ implies that $a \le \sqrt{x}$ or $b \le \sqrt{x}$. This allows him to write

$$\sum_{ab\le x} 1 = \sum_{a\le\sqrt{x}} \sum_{b\le x/a} 1 + \sum_{b\le\sqrt{x}} \sum_{a\le x/b} 1 - \sum_{a\le\sqrt{x}} \sum_{b\le\sqrt{x}} 1.$$

The first two double sums coincide, both being equal to

$$\sum_{m\le\sqrt{x}} \lfloor x/m \rfloor = \sum_{m\le\sqrt{x}} \left(\frac{x}{m} + O(1)\right) = x \sum_{m\le\sqrt{x}} \frac{1}{m} + O(\sqrt{x})$$

$$= x(\log\sqrt{x} + \gamma + O(1/\sqrt{x})) + O(\sqrt{x})$$

$$= \frac{1}{2} x \log x + \gamma x + O(\sqrt{x}).$$

(To go from the first line to the second, we used again the result of Problem 1.8.) The third double sum is of size $\lfloor\sqrt{x}\rfloor^2 = (\sqrt{x} + O(1))^2 = x + O(\sqrt{x})$. Collecting our estimates,

$$\sum_{n\le x} \tau(n) = x \log x + (2\gamma - 1)x + O(\sqrt{x}).$$

Voronoi showed in 1903 that the error term $O(\sqrt{x})$ in the last line can be replaced by $O(x^{1/3} \log(ex))$. A folklore conjecture claims that this error is in fact $O_\epsilon(x^{1/4+\epsilon})$, for any fixed $\epsilon > 0$. It is known (Hardy) that the "$+\epsilon$" is necessary, in the sense that the error is *not* $O(x^{1/4})$.

3.31

(a) This is easy: $\tau(n) = k + 1 > k > k \log 2 = \log n$.
(b) Fix a positive integer $m > A$, and fix m distinct primes $p_1, \ldots, p_m$. For each positive integer k,

$$\tau((p_1 \cdots p_m)^k) = (k + 1)^m.$$

On the other hand, for $C := (\log(p_1 \cdots p_m))^A$,

$$(\log((p_1 \cdots p_m)^k))^A = Ck^A.$$

Since m is larger than A, the quantity $(k + 1)^m$ will exceed Ck^A for all large enough values of k. So any n of the form $(p_1 \cdots p_m)^k$, with k large enough, satisfies the conclusion of (b).

3.32 The map $d \mapsto n/d$ sets up a one-to-one correspondence between the divisors of n below $\sqrt{n}$ and those above $\sqrt{n}$. It follows that

$$\tau(n) = 2 \sum_{\substack{d\mid n \\ d\le\sqrt{n}}} 1 - \begin{cases} 1 & \text{if } n \text{ is a square,} \\ 0 & \text{otherwise.} \end{cases}$$

The desired inequality $\tau(n) \le 2\sqrt{n}$ is an immediate consequence.

Remark. It is natural to ask about the optimality of the exponent $\frac{1}{2}$ in the bound $\tau(n) \le 2n^{1/2}$. Is $\tau(n) \le Cn^{1/3}$ (for some C and all n)? If so, what about $\tau(n) \le C'n^{1/100}$?

In fact, for each fixed $\epsilon > 0$, there is a C_ϵ such that

$$\tau(n) \le C_\epsilon n^\epsilon \tag{3.11}$$

for all $n \in \mathbb{Z}^+$. This result was first proved by Runge in 1885, who used it to show that, in a precise sense, almost all quintic polynomials $x^5 + ux + v$ ($u, v \in \mathbb{Z}$) are not solvable by radicals. A stronger result than (3.11) will be proved later (Problem 11.114), so we only give the idea of the proof here, leaving the reader to work out the details. First, one shows that if $p > 2^{1/\epsilon}$, then $\tau(p^k) = k+1 < p^{k\epsilon}$ for every positive integer k. Next, one argues that for any fixed p, the function $\tau(p^k)/p^{k\epsilon}$ assumes a maximum value as k ranges over $\mathbb{Z}^+$. Call this value $C_{p,\epsilon}$. As noted already, $C_{p,\epsilon} < 1$ when $p > 2^{1/\epsilon}$. On the other hand, it is easy to see (consider $k = 1$) that $C_{p,\epsilon} \ge 1$ for $p \le 2^{1/\epsilon}$. Thus, for each positive integer n,

$$\frac{\tau(n)}{n^\epsilon} \le \prod_{\substack{p^k\|n \\ p\le 2^{1/\epsilon}}} \frac{\tau(p^k)}{p^{k\epsilon}} \le \prod_{\substack{p\mid n \\ p\le 2^{1/\epsilon}}} C_{p,\epsilon} \le \prod_{p\le 2^{1/\epsilon}} C_{p,\epsilon}.$$

So we can take $C = \prod_{p\le 2^{1/\epsilon}} C_{p,\epsilon}$.

3.33 This is equivalent to showing that for each N, there is a positive integer d that occurs at least N times as a difference of primes $q - p$.

With x a large real number to be specified shortly, we consider all $\binom{\pi(x)}{2}$ pairs of primes (p, q) with $p < q \le x$. Since each difference $q - p < x$, there are fewer than x possible values for the difference $q - p$. Hence some difference occurs for at least

$$\frac{1}{x}\binom{\pi(x)}{2}$$

such pairs. We will argue that this quantity can be made sufficiently large by a suitable choice of x. From Problem 3.24, the sum of the reciprocals of the primes diverges. By Problem 3.28, there are arbitrarily large values of x with $\pi(x) > x/(\log x)^{1.01}$. For sufficiently large x of this kind, $\frac{1}{x}\binom{\pi(x)}{2} > x/(\log x)^{2.1}$ (say). This lower bound tends to infinity with x, and the result follows.

Remark. In 2013, Y. Zhang proved the breakthrough result that there are positive integers d that occur *infinitely often* as a difference $q - p$. The "twin prime conjecture", still open, makes the claim that $d = 2$ has this property.

Solutions to Step #4

4.34

(a) The first inequality is obvious, since $(x/p)^\epsilon \geq 1$ for every prime p over which the sum is taken. The second inequality then follows from Problem 3.24 with $s = 1 + \epsilon \, (\in (1, 2))$.

(b) We take $\epsilon = \frac{1}{\log x \cdot \log \log x}$, which belongs to $(0, 1)$ for large enough x. Since $\exp(u) = 1 + O(u)$ as $u \to 0$, we see that

$$x^\epsilon = \exp\left(\frac{1}{\log \log x}\right) = 1 + O\left(\frac{1}{\log \log x}\right).$$

Moreover,

$$\log(1/\epsilon) = \log(\log x \cdot \log \log x)$$

$$= \log \log x \left(1 + \frac{\log \log \log x}{\log \log x}\right).$$

Therefore,

$$x^\epsilon \log \frac{1}{\epsilon} = \log \log x \left(1 + \frac{\log \log \log x}{\log \log x}\right)\left(1 + O\left(\frac{1}{\log \log x}\right)\right)$$

$$= \log \log x + \log \log \log x + O(1).$$

So from part (a), keeping in mind that $x^\epsilon = O(1)$, we infer that

$$\sum_{p \leq x} \frac{1}{p} \leq \log \log x + \log \log \log x + O(1).$$

© Springer Nature Switzerland AG 2021
P. Pollack, A. Singha Roy, *Steps into Analytic Number Theory*, Problem Books in
Mathematics, https://doi.org/10.1007/978-3-030-65077-3_19

This is slightly sharper than the upper bound claimed in the problem statement.

Remark. How did we come up with this ϵ? We viewed x as fixed and chose ϵ to nearly minimize $x^\epsilon \log \frac{1}{\epsilon}$. Viewed as a function of ϵ, the quantity $x^\epsilon \log \frac{1}{\epsilon}$ blows up as $\epsilon \downarrow 0$ and as $\epsilon \to \infty$; being a continuously varying function of ϵ, it must therefore assume a global minimum on $(0, \infty)$. Taking derivatives, we see that the global minimum is attained at the unique solution ϵ to

$$\epsilon \log \epsilon = -\frac{1}{\log x}.$$

Equivalently, putting $t = \frac{1}{\epsilon}$, the minimum occurs when

$$t/\log t = \log x.$$

This last equation implicitly defines t as a function of x. While there is no simple closed form solution for t in terms of x, it is not hard to approximate t. Observe that as $x \to \infty$, so does our solution t. Moreover, taking logarithms,

$$\log t - \log \log t = \log \log x.$$

Thus,

$$\log t = \log \log x \left(1 - \frac{\log \log t}{\log t} \right)^{-1}.$$

Multiplying with $t/\log t = \log x$, we see that

$$t = \log x \cdot \log \log x \left(1 - \frac{\log \log t}{\log t} \right)^{-1}.$$

The parenthesized factor tends to 1 as x (and hence t) tends to infinity. Thus, $\log x \cdot \log \log x$ is a reasonable approximation to t for large x, making $\frac{1}{\log x \cdot \log \log x}$ a reasonable choice for ϵ.

4.35

(a) Clearly, $p^{1+\epsilon} > p$ for all primes p, so that taking $s = 1 + \epsilon$ in Problem 3.24 yields

$$\sum_{p \leq x} \frac{1}{p} \geq \sum_{p \leq x} \frac{1}{p^{1+\epsilon}} = \sum_p \frac{1}{p^{1+\epsilon}} - \sum_{p > x} \frac{1}{p^{1+\epsilon}} \geq \log \left(\frac{1}{\epsilon} \right) - \sum_{n > x} \frac{1}{n^{1+\epsilon}} + O(1).$$

(b) We insert into (a) the upper bound of Problem 2.11 for $\sum_{n > x} \frac{1}{n^{1+\epsilon}}$. This gives that

$$\sum_{p \leq x} \frac{1}{p} \geq \log \frac{1}{\epsilon} - \left(\frac{1+\epsilon}{\epsilon} \right) x^{-\epsilon} + O(1)$$

$$= \log \frac{1}{\epsilon} - \frac{x^{-\epsilon}}{\epsilon} + O(1).$$

Now we look to maximize the function $\log \frac{1}{\epsilon} - \frac{x^{-\epsilon}}{\epsilon}$ for $\epsilon \in (0, \infty)$. Taking derivatives, the global maximum occurs at an ϵ with

$$\frac{\log x}{\epsilon} + \frac{1}{\epsilon^2} = \frac{x^{\epsilon}}{\epsilon}.$$

Equivalently, writing $\epsilon = t / \log x$,

$$(\log x)^2 (t^{-1} + t^{-2}) = e^t (\log x) t^{-1},$$

or

$$(1 + t^{-1}) \log x = e^t.$$

The left side here is larger than $\log x$, and so for equality to hold, t must be large (certainly, $t > \log \log x$). Thus, $1 + t^{-1}$ is close to 1, and now referring back to the same equation, e^t is roughly equal to $\log x$. This suggests that the maximum is attained when $t \approx \log \log x$, so that $\epsilon \approx \log \log x / \log x$. And indeed if we take $\epsilon = \frac{\log \log x}{\log x}$, then

$$\log \frac{1}{\epsilon} = \log \log x - \log \log \log x,$$

while

$$\frac{1}{\epsilon} x^{-\epsilon} = \frac{\log x}{\log \log x} \cdot x^{-\frac{\log \log x}{\log x}} = \frac{\log x}{\log \log x} e^{-\log \log x} = \frac{1}{\log \log x} = O(1).$$

Thus, as $x \to \infty$,

$$\sum_{p \le x} \frac{1}{p} \ge \log \log x - \log \log \log x + O(1),$$

which is again slightly stronger than the claimed estimate.

Remark. Combining the results of Problems 4.34 and 4.35, we deduce that

$$\sum_{p \le x} \frac{1}{p} = \log \log x + O(\log \log \log x).$$

Later we will see how to establish still sharper estimates for $\sum_{p \le x} \frac{1}{p}$.

4.36 The identity $\phi(n) = n \sum_{d \mid n} \mu(d)/d = \sum_{d \mid n} \mu(d)(n/d)$ is of the form

$$f(n) = \sum_{d \mid n} g(d) h(n/d).$$

with $f(m) = \phi(m)$, $g(m) = \mu(m)$, and $h(m) = m$. So by Problem 3.29,

$$\sum_{n \le x} \phi(n) = \sum_{a \le x} \mu(a) \sum_{b \le x/a} b = \sum_{a \le x} \frac{\mu(a)}{2} \left(\left\lfloor \frac{x}{a} \right\rfloor + \left\lfloor \frac{x}{a} \right\rfloor^2 \right).$$

This last sum

$$= \frac{1}{2} \sum_{a \le x} \mu(a) \left\{ \left(\frac{x}{a} + O(1) \right) + \left(\frac{x}{a} + O(1) \right)^2 \right\}$$

$$= \frac{1}{2} \sum_{a \le x} \mu(a) \left\{ \left(\frac{x^2}{a^2} + O\left(\frac{x}{a}\right) \right) \right\}$$

$$= \frac{x^2}{2} \sum_{a \le x} \frac{\mu(a)}{a^2} + O\left(x \sum_{a \le x} \frac{1}{a} \right).$$

To finish, recall that $\sum_{a \le x} \frac{1}{a} = O(\log(ex))$ for all $x \ge 1$.

4.37 Let $x \ge 1$. By the triangle inequality and Problem 2.11,

$$\left| \sum_{a > x} \frac{\mu(a)}{a^2} \right| \le \sum_{a > x} \frac{1}{a^2} = O\left(\frac{1}{x} \right).$$

Now recalling the result of Problem 3.25,

$$\sum_{a \le x} \frac{\mu(a)}{a^2} = \sum_{a=1}^{\infty} \frac{\mu(a)}{a^2} - \sum_{a > x} \frac{\mu(a)}{a^2} = \frac{1}{\zeta(2)} + O\left(\frac{1}{x} \right).$$

To finish, insert this estimate into the result of Problem 4.36.

4.38 Say that $(a, b) \in \mathbb{Z}^2$ is a "coprime lattice point" if $\gcd(a, b) = 1$.

We start by considering those coprime lattice points in $(0, N] \times (0, N]$ that lie on or below the diagonal $y = x$. The number of these is

$$\sum_{a \le N} \#\{1 \le b \le a : \gcd(b, a) = 1\} = \sum_{a \le N} \phi(a).$$

The number of coprime lattice points on or above the diagonal is exactly the same, by the symmetry swapping a and b. Finally, there is precisely one coprime lattice point on the diagonal, namely $(1, 1)$. Hence, the total count of coprime lattice points in $(0, N] \times (0, N]$ is

$$2 \sum_{a \le N} \phi(a) - 1,$$

which, by the last problem, is

$$\frac{1}{\zeta(2)}N^2 + O(N \log (eN)).$$

To find the limiting probability, we divide by N^2 (the total number of lattice points in the square) and send $N \to \infty$; this gives $\frac{1}{\zeta(2)}$, as desired.

Remark. This result has an interesting geometric interpretation. It is easy to see that a lattice point P is coprime precisely when there is no lattice point on the interior of the segment OP, where $O = (0, 0)$ denotes the origin. For this reason, coprime lattice points are often called "visible points", meaning "visible from the origin". The result we just proved says (in rough terms) that a randomly chosen lattice point in the plane is visible with probability $\frac{1}{\zeta(2)}$.

It is natural to ask for the corresponding probability in higher dimensions. The general theorem is that a randomly chosen lattice point in $\mathbb{R}^n$ is coprime with probability $1/\zeta(n)$. How might one prove this? We have already handled the case $n = 2$, so assume that $n \geq 3$. Just as when $n = 2$,

$$(x_1, \ldots, x_n) \in \mathbb{Z}^n \text{ is visible} \iff \gcd(x_1, \ldots, x_n) = 1.$$

To detect the latter condition, we apply directly the following fundamental property of the Möbius function: If m is a positive integer, then $\sum_{d|m} \mu(d) = 1$ or 0, according to whether $m = 1$ or not. As a consequence, for any choice of integers $x_1, \ldots, x_n$, not all 0,

$$\sum_{d|x_1,\ldots,x_n} \mu(d) = \sum_{d|\gcd(x_1,\ldots,x_n)} \mu(d) = \begin{cases} 1 & \text{if } (x_1, \ldots, x_n) \text{ is visible,} \\ 0 & \text{otherwise.} \end{cases}$$

Hence, the total count of visible lattice points in $(0, N]^n$ is precisely

$$\sum_{x_1,\ldots,x_n \leq N} \sum_{d|x_1,\ldots,x_n} \mu(d) = \sum_{d \leq N} \mu(d)\lfloor N/d \rfloor^n.$$

We leave to the reader to work out that this last quantity has size $\frac{1}{\zeta(n)}N^n + O(N^{n-1})$, as $N \to \infty$.

There are many possible variations on this problem that are worth considering. For instance: What is the probability that a randomly chosen $(a_1, a_2, a_3) \in (\mathbb{Z}^+)^3$ is "pairwise coprime", meaning $\gcd(a_i, a_j) = 1$ for all pairs $i \neq j$? what is the analogous probability in $(\mathbb{Z}^+)^n$? Answers can be found in [1].

4.39 If $a \not\equiv 0 \pmod{m}$ then $e^{\frac{2\pi i a}{m}} \neq 1$. By the familiar formula for the partial sums of a geometric series,

$$\sum_{k \bmod m} e^{\frac{2\pi i k a}{m}} = \sum_{k=0}^{m-1}(e^{\frac{2\pi i a}{m}})^k = \frac{(e^{\frac{2\pi i a}{m}})^m - 1}{e^{\frac{2\pi i a}{m}} - 1} = 0.$$

On the other hand, if $a \equiv 0 \pmod{m}$, then the sum $\sum_{k \bmod m} e^{\frac{2\pi i k a}{m}}$ has all its terms equal to 1, and the result is immediate.

4.40 The identity in the previous problem tells us that $\frac{1}{m}\sum_{k \bmod m} e^{\frac{2\pi i k a}{m}}$ is the indicator function for the property "m divides a" (that is, it takes the value 1 when $m \mid a$, and takes the value 0 otherwise). Hence, for a given positive integer m, and given integers a and n,

$$\frac{1}{m} \sum_{k \bmod m} e^{\frac{2\pi i k(n-a^2)}{m}} = \begin{cases} 1 & \text{if } a^2 \equiv n \pmod{m}, \\ 0 & \text{otherwise.} \end{cases}$$

Summing on a, we deduce that the number of square roots of n modulo m is given by

$$\frac{1}{m} \sum_{a \bmod m} \sum_{k \bmod m} e^{\frac{2\pi i k(n-a^2)}{m}} = \frac{1}{m} \sum_{k \bmod m} e^{\frac{2\pi i k n}{m}} \sum_{a \bmod m} e^{\frac{-2\pi i k a^2}{m}},$$

exactly as claimed.

4.41 We begin with some general considerations. Suppose that $\sum_{n=1}^{\infty} a_n$ is any convergent series of complex numbers. Then $\lim_{n\to\infty} a_n = 0$ (the nth term test). Thus there is positive integer N such that $|a_n| \leq 1$ whenever $n > N$. Then $|a_n| \leq C$ for all natural numbers n if we set

$$C := \max\{1, |a_1|, \ldots, |a_N|\}.$$

Applying this general observation with $a_n = \frac{a(n)}{n^{s_0}}$, we get the first assertion of the problem. The next assertion then follows from the triangle inequality, the comparison test, and the convergence of $\sum_{n=1}^{\infty} \frac{1}{n^{s-s_0}}$ for all $s > s_0 + 1$.

It remains to prove the final claim, giving a limit expression for $a(m)$ for every natural number m. The claim can be rephrased as the assertion that

$$\lim_{s\to\infty} m^s \sum_{n>m} \frac{a(n)}{n^s} = 0.$$

We know already that $|a(n)| \leq Cn^{s_0}$ for every natural number n. So by the triangle inequality, for every $s > s_0 + 1$,

$$0 \leq \left| m^s \sum_{n>m} \frac{a(n)}{n^s} \right| \leq Cm^s \sum_{n>m} \frac{1}{n^{s-s_0}}.$$

Since $\frac{1}{t^{s-s_0}}$ is decreasing for $t > 0$,

$$\sum_{n>m} \frac{1}{n^{s-s_0}} \leq \int_m^{\infty} \frac{dt}{t^{s-s_0}} = \frac{1}{s-s_0-1} m^{s_0-s+1},$$

and

$$Cm^s \sum_{n>m} \frac{1}{n^{s-s_0}} \le \frac{C}{s-s_0-1} m^{s_0+1}.$$

The right-hand side clearly tends to 0 as $s \to \infty$, and the claim follows.

4.42 Every perfect power N can be written uniquely in the form n^m, where $n, m \ge 2$ and n is not a perfect power. (Here m may be determined as the gcd of the exponents in the prime factorization of N.) Therefore, using $'$ for a sum over non-perfect powers,

$$\sum_{N \text{ perfect power}} \frac{1}{N-1} = {\sum_{n\ge 2}}' \sum_{m\ge 2} \frac{1}{n^m-1} = {\sum_{n\ge 2}}' \sum_{m\ge 2} \frac{1/n^m}{1-1/n^m}$$

$$= {\sum_{n\ge 2}}' \sum_{m\ge 2} \sum_{k\ge 1} \frac{1}{n^{km}} = {\sum_{n\ge 2}}' \sum_{k\ge 1} \sum_{m>2} \frac{1}{n^{km}} = {\sum_{n\ge 2}}' \sum_{k\ge 1} \frac{1}{n^k(n^k-1)}.$$

Every integer from 2 onward has a unique representation in the form n^k, where $n \ge 2$, n is not a perfect power, and $k \ge 1$. Hence,

$${\sum_{n\ge 2}}' \sum_{k\ge 1} \frac{1}{n^k(n^k-1)} = \sum_{r\ge 2} \frac{1}{r(r-1)} = 1.$$

Here our carefree approach to series manipulations is justified by our dealing only with series of positive terms.

Reference

1. László Tóth, *The probability that k positive integers are pairwise relatively prime*. Fibonacci Quart. 40 (2002), 13–18.

Solutions to Step #5

5.43 Suppose first that $\left(\frac{k}{p}\right) = 1$. Then each nonzero square modulo p occurs twice as a value of $ka^2 \bmod p$, as a runs mod p, while 0 occurs once. This description of the values assumed by ka^2 is independent of k, and so the value of $\sum_{a \bmod p} e^{2\pi i ka^2/p}$ is the same for all k with $\left(\frac{k}{p}\right) = 1$. In particular, since 1 is a square,

$$\sum_{a \bmod p} e^{2\pi i ka^2/p} = \sum_{a \bmod p} e^{2\pi i \cdot 1 \cdot a^2/p} = G = \left(\frac{k}{p}\right)G$$

in this case. Now suppose that $\left(\frac{k}{p}\right) = -1$. Then every residue class appears twice when we take the concatenation of the lists $\{a^2\}$ and $\{ka^2\}$, where again a runs mod p. Thus,

$$\sum_{a \bmod p} e^{2\pi i a^2/p} + \sum_{a \bmod p} e^{2\pi i ka^2/p} = 2 \sum_{b \bmod p} e^{2\pi i b/p} = 0.$$

(Here the final relation is from Problem 4.39.) Hence,

$$\sum_{a \bmod p} e^{2\pi i ka^2/p} = -\sum_{a \bmod p} e^{2\pi i a^2/p} = -G = \left(\frac{k}{p}\right)G$$

in this case as well.

5.44 By Problem 4.40, the number of square roots of n modulo p is

$$\frac{1}{p} \sum_{k \bmod p} e^{2\pi i kn/p} \sum_{a \bmod p} e^{-2\pi i ka^2/p}.$$

© Springer Nature Switzerland AG 2021
P. Pollack, A. Singha Roy, *Steps into Analytic Number Theory*, Problem Books in
Mathematics, https://doi.org/10.1007/978-3-030-65077-3_20

We isolate the contribution of the terms with $k \equiv 0$ and then insert the result of the last problem:

$$\frac{1}{p} \sum_{k \bmod p} e^{2\pi ikn/p} \sum_{a \bmod p} e^{-2\pi ika^2/p}$$

$$= \frac{1}{p}\left(p + \sum_{\substack{k \bmod p \\ k \not\equiv 0}} e^{2\pi ikn/p} \sum_{a \bmod p} e^{-2\pi ika^2/p} \right)$$

$$= 1 + \frac{G}{p} \sum_{\substack{k \bmod p \\ k \not\equiv 0}} \left(\frac{-k}{p}\right) e^{2\pi ikn/p}.$$

The value of the sum on k is unchanged if we remove the restriction that $k \not\equiv 0$, and the first assertion of the problem follows. The expression for $\left(\frac{n}{p}\right)$ is a simple consequence, since the number of square roots of n modulo p is also given by the formula $1 + \left(\frac{n}{p}\right)$.

5.45 Starting from the identity of Problem 5.44 for $\left(\frac{n}{p}\right)$ and taking the squared modulus of both sides, we find that

$$\left(\frac{n}{p}\right)^2 = \frac{G \cdot \bar{G}}{p^2} \left(\sum_{k \bmod p} \left(\frac{-k}{p}\right) e^{2\pi ikn/p} \right) \left(\sum_{k' \bmod p} \left(\frac{-k'}{p}\right) e^{-2\pi ik'n/p} \right).$$

Summing both sides over $n \bmod p$ yields

$$p - 1 = \frac{G \cdot \bar{G}}{p^2} \sum_{k,k' \bmod p} \left(\frac{kk'}{p}\right) \sum_{n \bmod p} e^{2\pi i(k-k')n/p}$$

$$= \frac{G \cdot \bar{G}}{p} \sum_{k \bmod p} \left(\frac{k^2}{p}\right)$$

$$= \frac{G \cdot \bar{G}}{p}(p - 1).$$

(Here we used Problem 4.39 to evaluate the sum on n in the first line.) Rearranging, $G \cdot \bar{G} = p$.

Remark. Gauss showed in 1805 that $G = \sqrt{p}$ when $p \equiv 1$ (mod 4) and $G = i\sqrt{p}$ when $p \equiv 3$ (mod 4). There are by now several proofs of this result (see Chapter VI of [1] for four different arguments), but none that can really be called easy. We sketch one of the simpler arguments, due to Schur. The reader familiar with linear algebra is invited to fill in the details, consulting [1] for hints when necessary.

We let $\omega = e^{2\pi i/p}$, and we introduce the $p \times p$ matrix

$$A = (\omega^{ij})_{0 \le i,j < p} = \begin{pmatrix} 1 & 1 & 1 & \cdots & 1 \\ 1 & \omega & \omega^2 & \cdots & \omega^{p-1} \\ 1 & \omega^2 & \omega^4 & \cdots & \omega^{2(p-1)} \\ \vdots & \vdots & \vdots & \vdots & \vdots \\ 1 & \omega^{(p-1)} & \omega^{2(p-1)} & \cdots & \omega^{(p-1)^2} \end{pmatrix}.$$

Then

$$\mathrm{tr}(A) = \sum_{i=0}^{p-1} \omega^{i^2} = G.$$

Let $\lambda_1, \ldots, \lambda_p$ be the eigenvalues of A, listed with their (algebraic) multiplicities. Schur's proof runs as follows:

(1) One computes that $(A^2)_{ij} = 0$ or p, the latter occurring if and only if $p \mid i+j$. From this, one infers that $A^4 = p^2 I$ (with I the $p \times p$ identity matrix), and hence that each eigenvalue of A is a 4th root of p^2, and thus one of $\pm i\sqrt{p}$.

For $r = 0, 1, 2$, and 3, we let n_r denote the number of terms in the list $\lambda_1, \ldots, \lambda_p$ equal to $i^r \sqrt{p}$. Using that the trace of A is the sum of its eigenvalues,

$$G = (n_0 - n_2)\sqrt{p} + (n_1 - n_3)i\sqrt{p}. \tag{5.12}$$

(2) From (1), one sees easily that $\mathrm{tr}(A^2) = p$. The list of eigenvalues of A^2 is $\lambda_1^2, \ldots, \lambda_p^2$, and so $p = \mathrm{tr}(A^2) = \sum_i \lambda_i^2 = (n_0 + n_2)p - (n_1 + n_3)p$. Hence,

$$(n_0 + n_2) - (n_1 + n_3) = 1. \tag{5.13}$$

(3) A is a Vandermonde matrix, and so we can easily evaluate its determinant: $\det(A) = \prod_{0 \le i < j < p}(\omega^j - \omega^i)$. Wrestling with this expression for a bit, one deduces that $\det(A) \in i^{p(p-1)/2}\mathbb{R}^+$. On the other hand, writing the determinant as the product of eigenvalues, $\det(A) \in i^{n_1 + 2n_2 + 3n_3}\mathbb{R}^+$. Comparing these two results yields

$$\frac{p(p-1)}{2} \equiv n_1 + 2n_2 + 3n_3 \pmod 4. \tag{5.14}$$

With (5.12)–(5.14) available, the sign of the Gauss sum is within reach. Consider the case when $p \equiv 1 \pmod 4$. Then $G = \pm\sqrt{p}$, so from (5.12),

$$n_1 = n_3$$

and

$$n_0 - n_2 = \delta \in \{\pm 1\}.$$

Our task is to show that $\delta = 1$. From the last two displayed identities and (5.13), $2n_0 - \delta - 2n_1 = 1$, and thus

$$n_0 - n_1 = \frac{1 + \delta}{2}.$$

Therefore,

$$p = n_0 + n_1 + n_2 + n_3$$

$$= n_0 + \left(n_0 - \frac{1+\delta}{2}\right) + (n_0 - \delta) + \left(n_0 - \frac{1+\delta}{2}\right)$$

$$= 4n_0 - (1 + 2\delta).$$

Hence, working modulo 4,

$$\frac{p(p-1)}{2} \equiv \frac{p-1}{2} \equiv 2n_0 - 1 - \delta.$$

So by (5.14),

$$2n_0 - 1 - \delta \equiv n_1 + 2n_2 + 3n_3$$

$$\equiv n_1 + 2n_2 + 3n_1 \equiv 2n_2 \equiv 2(n_0 - \delta) \equiv 2n_0 - 2\delta,$$

giving $\delta \equiv 1 \pmod 4$. Since $\delta = \pm 1$, it must be that $\delta = 1$.

5.46 Suppose that $p_1 < \cdots < p_j$ are the primes not exceeding x. By the multinomial theorem,

$$\left(\sum_{p \leq x} \frac{1}{p}\right)^k = \sum_{\substack{e_1,\ldots,e_j \geq 0 \\ e_1 + \cdots + e_j = k}} \frac{k!}{e_1! e_2! \cdots e_j!} \cdot \frac{1}{p_1^{e_1} \cdots p_j^{e_j}}.$$

We view this as a weighted sum of the numbers $1/n$, where $n = p_1^{e_1} \cdots p_j^{e_j}$. Here the n that appear are those that factor into primes not exceeding x and satisfy $\Omega(n) = k$. In particular, every squarefree number $n \leq x$ with $\omega(n) = k$ appears. For each of those, the weight is $\frac{k!}{1! 1! \cdots 1!} = k!$, and the result follows.

5.47 We take the inequality established in the last problem and sum on k:

$$\sum_{\substack{n \leq x \\ n \text{ squarefree}}} \frac{1}{n} = \sum_{k \geq 0} \sum_{\substack{n \leq x \\ n \text{ squarefree} \\ \omega(n) = k}} \frac{1}{n} \leq \sum_{k \geq 0} \frac{1}{k!} \left(\sum_{p \leq x} \frac{1}{p}\right)^k = \exp\left(\sum_{p \leq x} \frac{1}{p}\right),$$

which proves the first assertion.

To prove the second, recall that every positive integer can be written as the product of a square and a squarefree number. Thus,

$$\zeta(2) \sum_{\substack{n \leq x \\ n \text{ squarefree}}} \frac{1}{n} = \sum_{\substack{m \geq 1 \\ n \leq x \\ n \text{ squarefree}}} \frac{1}{m^2 n} \geq \sum_{r \leq x} \frac{1}{r}.$$

Moreover,

$$\sum_{r \le x} \frac{1}{r} \ge \sum_{r \le x} \int_r^{r+1} \frac{dt}{t} = \log(\lfloor x \rfloor + 1) > \log x.$$

Stitching these results together,

$$\exp\left(\sum_{p \le x} \frac{1}{p}\right) \ge \sum_{\substack{n \le x \\ n \text{ squarefree}}} \frac{1}{n} \ge \frac{1}{\zeta(2)} \sum_{n \le x} \frac{1}{n} > \frac{\log x}{\zeta(2)} > \frac{\log x}{2}.$$

(In the last step we used that $\zeta(2) = 1 + \sum_{n \ge 2} \frac{1}{n^2} < 1 + \int_1^\infty \frac{dt}{t^2} = 2$.) Now taking logarithms,

$$\sum_{p \le x} \frac{1}{p} > \log \log x - \log 2,$$

which is slightly sharper than the inequality that was claimed.

5.48 We can write $\sigma(n) = \sum_{d \mid n} g(d) h(n/d)$, where g and h are the arithmetic functions defined by setting $g(m) = m$ and $h(m) = 1$ for all $m \in \mathbb{Z}^+$. By Problem 3.29,

$$\sum_{n \le x} \sigma(n) = \sum_{b \le x} \sum_{a \le x/b} a = \frac{1}{2} \sum_{b \le x} \lfloor x/b \rfloor (\lfloor x/b \rfloor + 1)$$

$$= \frac{1}{2} \sum_{b \le x} ((x/b)^2 + O(x/b)) = \frac{1}{2} x^2 \sum_{b \le x} \frac{1}{b^2} + O\left(x \sum_{b \le x} \frac{1}{b}\right).$$

The O-term here is $O(x \log(ex))$, while

$$\sum_{b \le x} \frac{1}{b^2} = \zeta(2) - \sum_{b > x} \frac{1}{b^2} = \zeta(2) + O\left(\frac{1}{x}\right).$$

Substituting these estimates above gives the desired result.

5.49 Since each of $2^{\omega(n)}$, $\tau(n)$, and $2^{\Omega(n)}$ is multiplicative, we can assume that n is a prime power, say $n = p^k$. In that case, the claim is that $2 \le k + 1 \le 2^k$. The first inequality is trivial, while the second is easily proved by induction on k.

5.50 Let $g_k(n)$ denote the number of representations of n as a product of k integers > 1, where order matters. Clearly,

$$g(n) = \sum_{k \ge 0} g_k(n).$$

Now notice that for every integer $k \geq 0$,

$$(\zeta(s) - 1)^k = \left(\sum_{m \geq 2} \frac{1}{m^s} \right)^k = \sum_{m_1, \ldots, m_k \geq 2} \frac{1}{(m_1 \cdots m_k)^s} = \sum_{N=1}^{\infty} \frac{g_k(N)}{N^s},$$

so that summing on k,

$$\sum_{k \geq 0} (\zeta(s) - 1)^k = \sum_{N=1}^{\infty} \frac{1}{N^s} \sum_{k \geq 0} g_k(N) = \sum_{N=1}^{\infty} \frac{g(N)}{N^s}.$$

We have so far ignored issues of convergence. This is not as much of a problem as it may first appear. Since all of our series manipulations have been on series of nonnegative terms, all of the identities claimed above hold without any restriction on s, provided equality is interpreted in the following sense: If one side is finite, the other is the same finite real number, while if one side is infinite, then both are infinite.

If we suppose now that $s > \rho$, then $1 < \zeta(s) < 2$, so that $0 < \zeta(s) - 1 < 1$. Thus,

$$\sum_{k \geq 0} (\zeta(s) - 1)^k = \frac{1}{1 - (\zeta(s) - 1)} = \frac{1}{2 - \zeta(s)},$$

and we obtain the identity claimed in the problem.

5.51 First notice that $a(1) = b(1)$, since by Problem 4.41,

$$a(1) = \lim_{s \to \infty} \sum_{n=1}^{\infty} \frac{a(n)}{n^s} = \lim_{s \to \infty} \sum_{n=1}^{\infty} \frac{b(n)}{n^s} = b(1).$$

Since $a(1) = b(1)$, we can remove the terms $\frac{a(1)}{1^s}$ and $\frac{b(1)}{1^s}$ from our starting Dirichlet series and deduce that

$$\sum_{n=2}^{\infty} \frac{a(n)}{n^s} = \sum_{n=2}^{\infty} \frac{b(n)}{n^s} \tag{5.15}$$

for all large enough real s. Multiplying both sides by 2^s and sending $s \to \infty$ gives—again using Problem 4.41—that $a(2) = b(2)$. Now remove the terms $a(2)/2^s$ and $b(2)/2^s$ from (5.15), multiply by 3^s and send $s \to \infty$ to get $a(3) = b(3)$. Continuing in this way, we obtain (by induction) that $a(n) = b(n)$ for all positive integers n.

5.52 For every integer $m \geq 2$,

$$\int_0^{\pi/2} \sin^m x \, dx = \int_0^{\pi/2} (\sin^{m-2} x)(1 - \cos^2 x) \, dx$$

$$= \int_0^{\pi/2} \sin^{m-2} x \, dx - \int_0^{\pi/2} \cos x \cdot (\sin^{m-2} x \cos x) \, dx$$

$$= \int_0^{\pi/2} \sin^{m-2} x \, dx - \int_0^{\pi/2} \cos x \cdot d\left(\frac{1}{m-1} \sin^{m-1} x\right).$$

Now integrating by parts,

$$\int_0^{\pi/2} \cos x \cdot d\left(\frac{1}{m-1} \sin^{m-1} x\right)$$

$$= \frac{1}{m-1} \cos x \sin^{m-1} x \Big|_0^{\pi/2} + \frac{1}{m-1} \int_0^{\pi/2} \sin^m x \, dx$$

$$= \frac{1}{m-1} \int_0^{\pi/2} \sin^m x \, dx.$$

Substituting this back above and rearranging,

$$\frac{m}{m-1} \int_0^{\pi/2} \sin^m x \, dx = \int_0^{\pi/2} \sin^{m-2} x \, dx,$$

and so

$$\int_0^{\pi/2} \sin^m x \, dx = \frac{m-1}{m} \int_0^{\pi/2} \sin^{m-2} x \, dx.$$

Both of the claimed identities now follow by induction, starting from

$$\int_0^{\pi/2} \sin^0 x \, dx = \frac{\pi}{2}$$

and

$$\int_0^{\pi/2} \sin x \, dx = 1.$$

5.53 For brevity, write I_m for $\int_0^{\pi/2} \sin^m x \, dx$. For x from the interval $[0, \pi/2]$, $0 \leq \sin x \leq 1$, so that $\sin^{m+1} x \leq \sin^m x$. It follows immediately that $I_{m+1} \leq I_m$ for all m. Consequently,

$$1 = \frac{I_{2n+1}}{I_{2n+1}} \leq \frac{I_{2n}}{I_{2n+1}} \leq \frac{I_{2n-1}}{I_{2n+1}} = \frac{2n+1}{2n}$$

for every positive integer n. By the squeeze theorem, $I_{2n}/I_{2n+1} \to 1$.
  From Problem 5.52,

$$\frac{I_{2n}}{I_{2n+1}} = \frac{\pi}{2} \cdot (2n+1) \cdot \prod_{i=1}^{n} \left(\frac{2i-1}{2i} \right)^2 = \frac{\pi}{2} \prod_{j=1}^{2n} \left(\frac{j+1}{j} \right) \prod_{i=1}^{n} \left(\frac{2i-1}{2i} \right)^2$$

$$= \frac{\pi}{2} \prod_{i=1}^{n} \left(\frac{(2i)(2i+1)}{(2i-1)(2i)} \right) \prod_{i=1}^{n} \left(\frac{2i-1}{2i} \right)^2 = \frac{\pi}{2} \prod_{i=1}^{n} \left(\frac{2i-1}{2i} \cdot \frac{2i+1}{2i} \right).$$

Hence,

$$\prod_{i=1}^{n} \left(\frac{2i}{2i-1} \cdot \frac{2i}{2i+1} \right) = \frac{\pi}{2} \cdot \left(\frac{I_{2n}}{I_{2n+1}} \right)^{-1}.$$

Send n to infinity to conclude that

$$\prod_{i=1}^{\infty} \left(\frac{2i}{2i-1} \cdot \frac{2i}{2i+1} \right) = \frac{\pi}{2},$$

as desired.

Reference

1. Edmund Landau, *Elementary number theory*. Chelsea Publishing Co., New York, N.Y., 1958.

Solutions to Step #6

6.54 Summing the geometric series,

$$\sum_{n \le N} e^{2\pi i k n/p} = \frac{e^{2\pi i k(N+1)/p} - e^{2\pi i k/p}}{e^{2\pi i k/p} - 1}$$

$$= \frac{e^{2\pi i k/p} \cdot e^{\pi i k N/p}}{e^{\pi i k/p}} \cdot \frac{e^{\pi i k N/p} - e^{-\pi i k N/p}}{e^{\pi i k/p} - e^{-\pi i k/p}}.$$

The first right-hand fraction has absolute value 1. The second has numerator $2i \sin(\pi k N/p)$ and denominator $2i \sin(\pi k/p)$. Thus,

$$\left| \sum_{n \le N} e^{2\pi i k n/p} \right| = \frac{|\sin \frac{\pi k N}{p}|}{|\sin \frac{\pi k}{p}|} \le \frac{1}{|\sin \frac{\pi k}{p}|},$$

as desired.

6.55 We begin by establishing the inequalities indicated in the hint.

We could prove (i) by mucking about with derivatives. But here is a slicker proof. Let $S(\theta) = \frac{\sin \pi \theta}{\pi \theta}$, with $S(0) = 1$. Then for all real θ,

$$S(\theta) = \int_0^1 \cos(\pi \theta t) \, dt.$$

Since $\cos(\pi u)$ is a decreasing function of u on $[0, \frac{1}{2}]$, we infer that whenever $0 \le \theta \le \theta' \le \frac{1}{2}$,

$$S(\theta) = \int_0^1 \cos(\pi \theta t) \, dt \ge \int_0^1 \cos(\pi \theta' t) \, dt = S(\theta').$$

© Springer Nature Switzerland AG 2021
P. Pollack, A. Singha Roy, *Steps into Analytic Number Theory*, Problem Books in Mathematics, https://doi.org/10.1007/978-3-030-65077-3_21

That is, S is also decreasing on $[0, \frac{1}{2}]$. In particular, $S(\theta) \geq S(\frac{1}{2}) = \frac{2}{\pi}$ for all $\theta \in [0, \frac{1}{2}]$. This gives (i) in the range $0 \leq \theta \leq \frac{1}{2}$; that (i) also holds when $-\frac{1}{2} \leq \theta \leq 0$ follows immediately upon noting that $S(\theta) = S(-\theta)$.

To prove (ii), observe that

$$
\int_{-1}^{1} \frac{1}{2n+t}\, dt = \int_{0}^{1} \left(\frac{1}{2n+t} + \frac{1}{2n-t} \right) dt
$$

$$
= \int_{0}^{1} \frac{4n}{4n^2-t^2}\, dt = \int_{0}^{1} \frac{1}{n - \frac{t^2}{4n}}\, dt > \int_{0}^{1} \frac{1}{n}\, dt = \frac{1}{n}.
$$

With these preliminaries out of the way, we can proceed with the main event, estimating $\sum_{n \leq N} \left(\frac{n}{p} \right)$. From Problem 5.44,

$$
\left(\frac{n}{p} \right) = \frac{G}{p} \sum_{k \bmod p} e^{2\pi i k n / p} \left(\frac{-k}{p} \right),
$$

where $G = \sum_{a \bmod p} e^{2\pi i a^2 / p}$ is the Gauss sum. Thus,

$$
\sum_{n \leq N} \left(\frac{n}{p} \right) = \frac{G}{p} \sum_{k \bmod p} \left(\frac{-k}{p} \right) \sum_{n \leq N} e^{2\pi i k n / p}.
$$

Recalling that $|G| = \sqrt{p}$ and keeping in mind the last problem,

$$
\left| \sum_{n \leq N} \left(\frac{n}{p} \right) \right| \leq \frac{\sqrt{p}}{p} \sum_{\substack{k \bmod p \\ k \neq 0}} \left| \sum_{n \leq N} e^{2\pi i k n / p} \right| \leq \frac{1}{\sqrt{p}} \sum_{\substack{k \in \mathbb{Z} \\ |k| < p/2,\ k \neq 0}} \frac{1}{|\sin(\frac{\pi k}{p})|}
$$

$$
\leq \sqrt{p} \sum_{\substack{|k| < p/2,\ k \neq 0}} \frac{1}{2|k|} = \sqrt{p} \sum_{k=1}^{(p-1)/2} \frac{1}{k}.
$$

(In going from the first line to the second, we used that $|\frac{k}{p}| \leq \frac{1}{2}$, so that $|\sin(\pi \frac{k}{p})| \geq 2\frac{|k|}{p}$ by (i).) This proves the first inequality. The second follows from (ii):

$$
\sum_{k=1}^{(p-1)/2} \frac{1}{k} < \sum_{k=1}^{(p-1)/2} \int_{-1}^{1} \frac{1}{2k+t}\, dt = \sum_{k=1}^{(p-1)/2} (\log(2k+1) - \log(2k-1)) = \log p.
$$

6.56 Let N be a positive integer with $N < p$. Let A be the number of integers in $[1, N]$ that are squares mod p, and let B be the number of integers in the same interval that are nonsquares mod p. Then

$$A + B = N \quad \text{while} \quad A - B = \sum_{n \le N} \left(\frac{n}{p} \right).$$

Thus, the proportion of integers in $[1, N]$ that are squares mod p is

$$\frac{A}{N} = \frac{1}{2} + \frac{1}{2N} \sum_{n \le N} \left(\frac{n}{p} \right).$$

By the Pólya–Vinogradov inequality, this is bounded below by

$$\frac{1}{2} - \frac{\sqrt{p} \log p}{2N}.$$

Now take $N = \lfloor p^{\frac{1}{2} + \epsilon} \rfloor$. Then the displayed quantity exceeds 0.49 for all large p. Hence, at least 49% of the integers in $[1, p^{\frac{1}{2} + \epsilon}]$ are squares mod p. The proof of the statement for nonsquares is entirely analogous: The proportion of nonsquares in $[1, N]$ is

$$\frac{B}{N} = \frac{1}{2} - \frac{1}{2N} \sum_{n \le N} \left(\frac{n}{p} \right),$$

and the Pólya–Vinogradov inequality finishes matters as before.

Remark. Let p be a prime and let χ_0 be a nontrivial group homomorphism from $\mathbb{U}_p$ to the unit circle (meaning the group of complex numbers with absolute value 1). Define a function $\chi : \mathbb{Z} \to \mathbb{C}$ by letting $\chi(n) = \chi_0(n \bmod p)$ if $p \nmid n$ and $\chi(n) = 0$ otherwise. Such a function is called a Dirichlet character mod p. (You will encounter general Dirichlet characters in the Special Steps.) One can show that $|\sum_{n \le N} \chi(n)| < \sqrt{p} \log p$ for any p and χ as above, and any positive integer N. Note that this generalizes the inequality proved in Problem 6.55, since $\left(\frac{\cdot}{p} \right) = \chi(\cdot)$ when p is odd and χ_0 is the nontrivial homomorphism from $\mathbb{U}_p$ to $\{\pm 1\}$. This general Pólya–Vinogradov inequality has a number of pleasant consequences. For example, it implies that for each $\epsilon > 0$ and all $p > p_0(\epsilon)$, there is an integer smaller than $p^{\frac{1}{2} + \epsilon}$ generating the cyclic group $\mathbb{U}_p$. For this and much more, see Chapter 8 of Shapiro's textbook [6].

6.57 The expression for $\pi(x, y)$ in terms of alternating sums follows from the Principle of Inclusion-Exclusion, by the same reasoning as in Problem 2.19.

Now we drop the floor functions. Replacing $\lfloor x \rfloor$ with x incurs an error of at most 1 (in absolute value). By the triangle inequality, replacing $\sum_{p_1 \le y} \lfloor \frac{x}{p_1} \rfloor$ with $\sum_{p_1 \le y} \frac{x}{p_1}$ incurs an error of size at most $\pi(y)$—the number of terms in the sum. In general, when we replace

$$\sum_{p_1 < \cdots < p_r \le x} \left\lfloor \frac{x}{p_1 \cdots p_r} \right\rfloor \quad \text{with} \quad \sum_{p_1 < \cdots < p_r \le x} \frac{x}{p_1 \cdots p_r},$$

we incur an error of size at most $\binom{\pi(y)}{r}$.

We conclude that $\pi(x, y)$ is equal to

$$x - \sum_{p_1 \leq y} \frac{x}{p_1} + \sum_{p_1 < p_2 \leq y} \frac{x}{p_1 p_2} - \sum_{p_1 < p_2 < p_3 \leq y} \frac{x}{p_1 p_2 p_3} + \dots, \tag{6.16}$$

up to an error bounded in absolute value by

$$\sum_{r=0}^{\pi(y)} \binom{\pi(y)}{r}. \tag{6.17}$$

Now (6.16) is precisely $x \prod_{p \leq y} (1 - 1/p)$, while (6.17) is precisely $2^{\pi(y)}$. This proves the claimed estimate for $\pi(x, y)$, with an implied O-constant of 1.

6.58 We first prove that $\exp(t) \geq 1 + t$ for all real numbers t. Let $\Delta(t) = \exp(t) - (1+t)$, and notice that $\Delta'(t) = \exp(t) - 1$, so that Δ is strictly decreasing on $(-\infty, 0]$ and strictly increasing on $[0, \infty)$. Since $\Delta(0) = 0$, we deduce that $\Delta(t) \geq 0$ for all t, with equality only when $t = 0$.

As an immediate consequence,

$$\prod_{p \leq y} \left(1 - \frac{1}{p}\right) \leq \exp\left(-\sum_{p \leq y} \frac{1}{p}\right).$$

By Problem 5.47,

$$\exp\left(-\sum_{p \leq y} \frac{1}{p}\right) \leq \exp\left(-\log\log y + 1\right) = e/\log y.$$

(In fact, the proof of Problem 5.47 allows us to take the numerator as $\zeta(2)$ rather than e. But this will not be important.)

Alternative Solution One can prove the slightly sharper estimate

$$\prod_{p \leq y} \left(1 - \frac{1}{p}\right) \leq \frac{1}{\log y}$$

by borrowing some ideas used in the solution to Problem 3.22. For every $y \geq 1$,

$$\prod_{p \leq y} \left(1 - \frac{1}{p}\right)^{-1} = \prod_{p \leq y} \sum_{e_p \geq 0} \frac{1}{p^{e_p}} = \sum_{m: \, p|m \Rightarrow p \leq y} \frac{1}{m}.$$

Here the final sum ranges over all positive integers m which factor into primes not exceeding y. Since every positive integer $m \leq y$ is of this kind,

$$\sum_{m:\ p|m \Rightarrow p \le y} \frac{1}{m} \ge \sum_{m \le y} \frac{1}{m} > \log y.$$

6.59 That $\pi(x, y) = O(x/\log y + 2^{\pi(y)})$ follows immediately from the last two problems. We now choose $y = \log x$, so that $x/\log y = x/\log \log x$. For this y, we have $2^{\pi(y)} \le 2^y = x^{\log 2}$. Since $x^{\log 2} = O(x/\log \log x)$ as $x \to \infty$ (in fact, $x^{\log 2}/(x/\log \log x) \to 0$ as $x \to \infty$), we conclude that

$$\pi(x, \log x) = O(x/\log \log x),$$

as $x \to \infty$.

To deduce a bound on $\pi(x)$ from this, observe that

$$\pi(x) \le \pi(x, y) + y$$

for every choice of $x, y \ge 2$. Indeed, if $p \le x$ is prime, then either $p \le y$, in which case it is can be thought of as counted by the term y on the right-hand side, or p has no prime factors up to y, in which case p is counted by the $\pi(x, y)$ term. Specializing again to $y = \log x$, we get from our work above that

$$\pi(x) = O(x/\log \log x + \log x).$$

Since $\log x = O(x/\log \log x)$, we get that $\pi(x) = O(x/\log \log x)$, as claimed.

The quantity $\log \log x \to \infty$ as $x \to \infty$, so we deduce that 0% of the positive integers are prime, in the following precise sense: An integer chosen uniformly at random from $[1, x]$ is prime with probability tending to 0, as $x \to \infty$.

6.60 With $S(0)$ understood to vanish,

$$\sum_{n \le x} a_n f(n) = \sum_{n \le x} (S(n) - S(n-1)) f(n)$$

$$= \sum_{n \le x} S(n) f(n) - \sum_{2 \le n \le x} S(n-1) f(n)$$

$$= S(\lfloor x \rfloor) f(\lfloor x \rfloor) + \sum_{n \le x-1} S(n) f(n) - \sum_{n \le x-1} S(n) f(n+1)$$

$$= S(\lfloor x \rfloor) f(\lfloor x \rfloor) + \sum_{n \le x-1} S(n)(f(n) - f(n+1)).$$

6.61 Since $S(t) = S(n)$ for all $t \in [n, n+1)$,

$$-\int_n^{n+1} S(t) f'(t)\, dt = -S(n) \int_n^{n+1} f'(t)\, dt = S(n)(f(n) - f(n+1)),$$

which is the first assertion. Combining this with the result of the last problem,

$$\sum_{n\leq x} a_n f(n) = S(x)f(\lfloor x \rfloor) + \sum_{n\leq x-1}\left(-\int_n^{n+1} S(t)f'(t)\,dt\right)$$

$$= S(x)f(\lfloor x \rfloor) - \int_1^{\lfloor x \rfloor} S(t)f'(t)\,dt.$$

When $x \in \mathbb{Z}$, this looks exactly how we want. In general, it differs from what we want by

$$S(x)(f(x) - f(\lfloor x \rfloor)) - \int_{\lfloor x \rfloor}^x S(t)f'(t)\,dt.$$

But this is easily seen to be 0! (Pull out $S(t)$ from the integral as $S(\lfloor x \rfloor)$, and proceed as at the start of this solution.)

6.62 Let $a_n = 1$ when n is prime and $a_n = 0$ otherwise, so that $S(x) = \pi(x)$ in the setup for summation by parts. Let $f(t) = 1/t$, so that $f'(t) = -1/t^2$. Then

$$\sum_{p\leq x}\frac{1}{p} = \sum_{n\leq x} a_n f(n) = \frac{\pi(x)}{x} + \int_1^x \frac{\pi(t)}{t^2}\,dt.$$

The integral over t can (and will, below) be taken to start at 2, since $\pi(t) = 0$ when $t < 2$.

We are supposing that $\frac{\pi(x)}{x/\log x} \to A$ (as $x \to \infty$), for some $A \in \mathbb{R}$. Thus, if we define $E(x)$ for $x \geq 2$ by the equation $\pi(x) = Ax/\log x + E(x)$, then $\frac{E(x)}{x/\log x} \to 0$. Also,

$$\sum_{p\leq x}\frac{1}{p} = \frac{\pi(x)}{x} + \int_2^x\left(\frac{A}{t\log t} + \frac{E(t)}{t^2}\right)dt$$

$$= A\log\log x + O(1) + \int_2^x \frac{E(t)}{t^2}\,dt.$$

When we divide by $\log\log x$ and send x to ∞, the first term on the right will tend to A while the second will tend to 0. So we will have proved that $\frac{\sum_{p\leq x} 1/p}{\log\log x} \to A$ once we have shown that

$$\frac{1}{\log\log x}\int_2^x \frac{E(t)}{t^2}\,dt \to 0. \tag{6.18}$$

To prove (6.18), we get our hands dirty and work with the limit definition. Fix $\epsilon > 0$. Since $\frac{E(x)}{x/\log x} \to 0$, we know that $|E(t)| < \epsilon\frac{t}{\log t}$ for all sufficiently large t. It will be convenient to have an upper bound valid for all t, not only those sufficiently large. Since $E(t)$ is bounded on every finite, closed subinterval of $[2, \infty)$, we can say that for some constant C (which depends on ϵ),

$$|E(t)| \leq \epsilon \frac{t}{\log t} + C \quad \text{for all } t \geq 2.$$

It follows that

$$\left| \int_2^x \frac{E(t)}{t^2} \, dt \right| \leq \int_2^x \left(\frac{\epsilon}{t \log t} + \frac{C}{t^2} \right) \leq \epsilon \log \log x + O(1).$$

In particular,

$$\frac{\left| \int_2^x E(t) t^{-2} \, dt \right|}{\log \log x} < 2\epsilon$$

for all large enough x. Since $\epsilon > 0$ was arbitrary, (6.18) follows.

Finally, we know from Step #4 that $\sum_{p \leq x} \frac{1}{p} = \log \log x + O(\log \log \log x)$, and so

$$\lim_{x \to \infty} \frac{\sum_{p \leq x} 1/p}{\log \log x} = 1.$$

We conclude that if $\frac{\pi(x)}{x/\log x} \to A$, then $A = 1$.

6.63 Taking $x = \sin t$, we see that for $0 \leq t \leq \pi/2$,

$$t = \arcsin(\sin t)$$

$$= \sum_{n=0}^{\infty} \frac{1 \cdot 3 \cdots (2n-1)}{2 \cdot 4 \cdots 2n} \frac{\sin^{2n+1} t}{2n+1}.$$

Thus, by Problem 5.52,

$$\frac{\pi^2}{8} = \int_0^{\pi/2} t \, dt$$

$$= \int_0^{\pi/2} \sum_{n=0}^{\infty} \frac{1 \cdot 3 \cdots (2n-1)}{2 \cdot 4 \cdots 2n} \frac{\sin^{2n+1} t}{2n+1} \, dt$$

$$= \sum_{n=0}^{\infty} \frac{1 \cdot 3 \cdots (2n-1)}{2 \cdot 4 \cdots 2n} \cdot \frac{1}{2n+1} \int_0^{\pi/2} \sin^{2n+1} t \, dt$$

$$= \sum_{n=0}^{\infty} \frac{1 \cdot 3 \cdots (2n-1)}{2 \cdot 4 \cdots 2n} \cdot \frac{1}{2n+1} \cdot \frac{2 \cdot 4 \cdots 2n}{3 \cdot 5 \cdots (2n+1)}$$

$$= \sum_{n=0}^{\infty} \frac{1}{(2n+1)^2}.$$

(The interchange of the sum and the integral when moving from line 2 to line 3 can be justified either (a) by the nonnegativity of the summands, via Fubini–Tonelli, or (b) by an argument involving uniform convergence.) This proves the first identity. The second is an easy consequence. Indeed,

$$\sum_{n=1}^{\infty} \frac{1}{n^2} = \sum_{n=1}^{\infty} \frac{1}{(2n)^2} + \sum_{n=0}^{\infty} \frac{1}{(2n+1)^2}$$

$$= \frac{1}{4} \sum_{n=1}^{\infty} \frac{1}{n^2} + \sum_{n=0}^{\infty} \frac{1}{(2n+1)^2},$$

so that

$$\sum_{n=1}^{\infty} \frac{1}{n^2} = \frac{4}{3} \sum_{n=0}^{\infty} \frac{1}{(2n+1)^2} = \frac{4}{3} \cdot \frac{\pi^2}{8} = \frac{\pi^2}{6}.$$

Remark. Euler was the first to show that $\zeta(2) = \frac{\pi^2}{6}$, in 1734. Today dozens of proofs are known. Several of these have been collected by Robin Chapman in [2]. The proof we have described is a variant of a lesser-known argument of Euler (see §7 of [1]). The precise version given here appears as a footnote in Knopp's book on infinite series (see p. 376 of [5]).

We cannot resist sketching the beautiful, deeply *arithmetic* proof of Loo Keng Hua (found on p. 209 of [3]). Hua's argument has as its starting point Jacobi's version of the four-square theorem. For each positive integer n, let

$$r(n) = \#\{(x, y, z, w) \in \mathbb{Z}^4 : x^2 + y^2 + z^2 + w^2 = n\}.$$

Refining Lagrange's theorem that $r(n) > 0$ for all positive integers n, Jacobi shows the exact formula

$$r(n) = 8\sigma(n) - 32\sigma(n/4), \tag{6.19}$$

where σ is the usual sum-of-divisors function, and where $\sigma(n/4)$ should be interpreted as 0 when $4 \nmid n$. A proof can be found in [3, Chapter 8] (alternatively, see [4, Chapter 17]).

Writing $S(x)$ for $\sum_{n \leq x} \sigma(n)$, we infer from (6.19) and Problem 5.48 that as $N \to \infty$,

$$\sum_{n \leq N} r(n) = 8S(N) - 32S(N/4)$$

$$= 3\zeta(2)N^2 + O(N \log N).$$

Now we look at the sum $\sum_{n \leq N} r(n)$ from a different perspective. This sum can be thought of geometrically: Notice that

$$\sum_{n \leq N} r(n) = \#\{(x, y, z, w) \in \mathbb{Z}^4 : 0 < x^2 + y^2 + z^2 + w^2 \leq N\},$$

and that the right-hand side counts the number of nonzero lattice points in the 4-dimensional ball of radius $\sqrt{N}$ about the origin. One might guess that this count is roughly the same as the 4-volume of the ball, $\frac{\pi^2}{2} N^2$. This guess can be validated by the method used in the solution to Problem 2.16; in fact, one can show in that way that the right-hand side here is

$$\frac{\pi^2}{2} N^2 + O(N^{3/2}),$$

as $N \to \infty$. Equating our two estimates for $\sum_{n \leq N} r(n)$, dividing by N^2, and sending $N \to \infty$, we conclude that $3\zeta(2) = \frac{\pi^2}{2}$. Thus, $\zeta(2) = \frac{\pi^2}{6}$. Q.E.D.

6.64 In the solution to Problem 5.50, we saw that

$$\sum_{n=1}^{\infty} \frac{g(n)}{n^s} = \sum_{k \geq 0} (\zeta(s) - 1)^k,$$

with both sides convergent for $s > \rho$. (Here ρ is the unique real number larger than 1 with $\zeta(\rho) = 2$.) By essentially the same argument,

$$\sum_{n=1}^{\infty} \frac{g_e(n)}{n^s} = \sum_{k \geq 0} (\zeta(s) - 1)^{2k}$$

and

$$\sum_{n=1}^{\infty} \frac{g_o(n)}{n^s} = \sum_{k \geq 0} (\zeta(s) - 1)^{2k+1},$$

with all the series convergent for $s > \rho$. Hence, for $s > \rho$,

$$\sum_{n=1}^{\infty} \frac{g_e(n) - g_o(n)}{n^s} = 1 - (\zeta(s) - 1) + (\zeta(s) - 1)^2 - \dots$$

$$= \frac{1}{1 + (\zeta(s) - 1)} = \frac{1}{\zeta(s)} = \sum_{n=1}^{\infty} \frac{\mu(n)}{n^s}.$$

(Here the final equality is from the solution to Problem 3.25.) Now we appeal to Problem 5.51: If two Dirichlet series represent the same function for all large enough s, then their coefficients are equal. Consequently, $g_e(n) - g_o(n) = \mu(n)$ for all positive integers n.

References

1. Raymond Ayoub, *Euler and the zeta function.* Amer. Math. Monthly 81 (1974), 1067–1086.
2. Robin J. Chapman, *Evaluating* $\zeta(2)$. Published online at https://empslocal.ex.ac.uk/people/staff/rjchapma/etc/zeta2.pdf.
3. Loo Keng Hua, *Introduction to number theory.* Translated from the Chinese by Peter Shiu. Springer-Verlag, Berlin-New York, 1982.
4. Kenneth Ireland and Michael A. Rosen, *A classical introduction to modern number theory.* Second edition. Graduate Texts in Mathematics, vol. 84. Springer-Verlag, New York, 1990.
5. Konrad Knopp, *Theory and application of infinite series.* Blackie and Son, London–Glasgow, 1954.
6. Harold N. Shapiro, *Introduction to the theory of numbers.* John Wiley & Sons, Inc., New York, 1983.

Solutions to Step #7

7.65 Let p be a prime dividing $\frac{\mathrm{LCM}(2n+1)}{\mathrm{LCM}(n+1)}$. We must show that $p^2 \nmid \frac{\mathrm{LCM}(2n+1)}{\mathrm{LCM}(n+1)}$.

Let p^k be the highest power of p dividing $\mathrm{LCM}(2n+1)$, so that $k \geq 1$. Then $p^k \leq 2n+1$ (in fact, p^k is the largest power of p with $p^k \leq 2n+1$). Thus, $p^{k-1} \leq \frac{1}{2}p^k < n+1$, and so $p^{k-1} \mid \mathrm{LCM}(n+1)$. As p^k is the highest power of p dividing $\mathrm{LCM}(2n+1)$ and $p^{k-1} \mid \mathrm{LCM}(n+1)$, we conclude that $p^2 \nmid \frac{\mathrm{LCM}(2n+1)}{\mathrm{LCM}(n+1)}$.

7.66 Since the ratio $\frac{\mathrm{LCM}(2n+1)}{\mathrm{LCM}(n+1)}$ is squarefree, it is enough to show that every prime dividing this ratio also divides $\binom{2n+1}{n+1}$. For this we will use the following well-known theorem of Legendre: If p is prime and $x \geq 1$, then the exponent on the highest power of p dividing $\lfloor x \rfloor!$ is

$$\left\lfloor \frac{x}{p} \right\rfloor + \left\lfloor \frac{x}{p^2} \right\rfloor + \left\lfloor \frac{x}{p^3} \right\rfloor + \cdots.$$

Let p be a prime dividing $\frac{\mathrm{LCM}(2n+1)}{\mathrm{LCM}(n+1)}$. Then the highest power of p less than or equal to $2n+1$ cannot also be less than or equal to $n+1$. Thus there is some power of p, say p^k, in the interval $(n+1, 2n+1]$. By Legendre's formula, the exponent on p in the factorization of $\binom{2n+1}{n+1} = \frac{(2n+1)!}{(n+1)!n!}$ is

$$\sum_{r \geq 1} \left(\left\lfloor \frac{2n+1}{p^r} \right\rfloor - \left\lfloor \frac{n+1}{p^r} \right\rfloor - \left\lfloor \frac{n}{p^r} \right\rfloor \right).$$

By the choice of k, the term of the sum corresponding to $r = k$ is positive. So to finish the proof, it is enough to show that every term in the sum is nonnegative. This is easy: First observe that

© Springer Nature Switzerland AG 2021
P. Pollack, A. Singha Roy, *Steps into Analytic Number Theory*, Problem Books in
Mathematics, https://doi.org/10.1007/978-3-030-65077-3_22

$$\left\lfloor \frac{n+1}{p^r} \right\rfloor + \left\lfloor \frac{n}{p^r} \right\rfloor \leq \frac{n+1}{p^r} + \frac{n}{p^r} = \frac{2n+1}{p^r}.$$

Since the left-hand side is an integer, the inequality $\left\lfloor \frac{n+1}{p^r} \right\rfloor + \left\lfloor \frac{n}{p^r} \right\rfloor \leq \left\lfloor \frac{2n+1}{p^r} \right\rfloor$ follows.

For completeness, we give a proof of Legendre's formula. For each positive integer m, the exponent on the highest of power of p dividing m can be written as $\sum_{j \geq 1:\ p^j \mid m} 1$. Hence, the exponent on p appearing in the factorization of $\lfloor x \rfloor! = \prod_{n \leq x} n$ is

$$\sum_{\substack{n \leq x \\ p^j \mid n}} \sum_{j \geq 1} 1 = \sum_{j \geq 1} \sum_{\substack{n \leq x \\ p^j \mid n}} 1 = \sum_{j \geq 1} \left\lfloor \frac{x}{p^j} \right\rfloor.$$

7.67 The first inequality is clear, since $\binom{2n+1}{n} = \binom{2n+1}{n+1}$, and both of those terms appear in $\sum_{k \geq 0} \binom{2n+1}{k}$. That $\frac{1}{2} \sum_{k \geq 0} \binom{2n+1}{k} = 4^n$ follows from the binomial theorem, noting that $\sum_{k \geq 0} \binom{2n+1}{k} = (1+1)^{2n+1}$.

7.68 If the proposed inequality fails, let n be the smallest counterexample.

If n is even, then $\frac{n}{2} \leq n - 1$. Thus, $n = 2 \cdot \frac{n}{2}$ divides $2 \cdot \text{LCM}(n-1)$. Clearly, each of $1, 2, \ldots, n-1$ also divides $2 \cdot \text{LCM}(n-1)$. Hence,

$$\text{LCM}(n) \leq 2 \cdot \text{LCM}(n-1) \leq 2 \cdot 4^{n-1} < 4^n,$$

contrary to hypothesis.

So n must be odd. Write $n = 2m + 1$, where $m \in \mathbb{Z}$. Clearly, $n > 1$, and so $m \in \mathbb{Z}^+$. From Problems 7.66 and 7.67,

$$\frac{\text{LCM}(2m+1)}{\text{LCM}(m+1)} \leq 4^m,$$

so that

$$\text{LCM}(n) = \text{LCM}(2m+1) \leq 4^m \cdot \text{LCM}(m+1) \leq 4^m \cdot 4^{m+1} = 4^{2m+1} = 4^n.$$

So n is not a counterexample to the inequality after all, a contradiction.

7.69 It is easy to prove (with calculus or without) that $0 \leq t(1-t) \leq \frac{1}{4}$ for all $t \in [0, 1]$. That

$$\frac{1}{4^n} \geq \int_0^1 t^n (1-t)^n \, dt.$$

follows immediately. If we expand out $t^n(1-t)^n$, we obtain a sum of terms At^m, where $A \in \mathbb{Z}$ and $n \leq m \leq 2n$. For each of these terms,

$$\text{LCM}(2n+1) \int_0^1 At^m \, dt = A \cdot \frac{\text{LCM}(2n+1)}{m+1} \in \mathbb{Z},$$

and so

$$\text{LCM}(2n+1) \int_0^1 t^n(1-t)^n \, dt \in \mathbb{Z}.$$

To replace $\mathbb{Z}$ with $\mathbb{Z}^+$ on the right-hand side, it remains only to note that the left-hand side is positive (which is clear, since $t^n(1-t)^n$ is continuous on $[0,1]$, nonnegative-valued, and not identically 0).

Since 1 is the least positive integer,

$$\text{LCM}(2n+1)\frac{1}{4^n} \geq \text{LCM}(2n+1) \int_0^1 t^n(1-t)^n \, dt \geq 1,$$

and so

$$\text{LCM}(2n+1) \geq 4^n.$$

7.70 Every block of m consecutive positive integers contains precisely one integer belonging to a given congruence class mod m. The set of positive integers $\leq x$ contains the first $\lfloor x/m \rfloor$ such blocks and is contained within the first $\lceil x/m \rceil$ such blocks. Both $\lfloor x/m \rfloor$ and $\lceil x/m \rceil$ are within 1 of x/m, and the result follows.

7.71 Let $\mathcal{A}_d$ denote the set of $n \leq x$ for which $d \mid n(n+2)$, so that $A_d = \#\mathcal{A}_d$. In this notation,

$$\pi_2(x,y) = \left| \mathcal{A}_1 \setminus \cup_{p \leq y} \mathcal{A}_p \right|.$$

Each $\mathcal{A}_p$ is a subset of $\mathcal{A}_1$, so by inclusion-exclusion, our desired count is

$$|\mathcal{A}_1| - \sum_{p_1 \leq y} |\mathcal{A}_{p_1}| + \sum_{p_1 < p_2 \leq y} |\mathcal{A}_{p_1} \cap \mathcal{A}_{p_2}| - \sum_{p_1 < p_2 < p_3 \leq y} |\mathcal{A}_{p_1} \cap \mathcal{A}_{p_2} \cap \mathcal{A}_{p_3}| + \ldots.$$

This gives the first formula for $\pi_2(x,y)$, once we recognize that $\mathcal{A}_{p_1} \cap \cdots \cap \mathcal{A}_{p_r} = \mathcal{A}_{p_1 \cdots p_r}$ for all choices of distinct primes $p_1, \ldots, p_r$.

The divisors of $P = \prod_{p \leq y}$ are precisely the numbers $p_1 p_2 \cdots p_r$, for all choices of primes $p_1 < \cdots < p_r \leq y$. The second formula now follows from the definition of the Möbius function.

7.72 We start by estimating A_d for each d dividing $P = \prod_{p \leq y} p$.

Every d dividing P is squarefree, and so

$$n(n+2) \equiv 0 \pmod{d} \iff n(n+2) \equiv 0 \pmod{p} \text{ for all primes } p \mid d.$$

Hence, by the Chinese remainder, if we let $v(d)$ denote the number of mod d solutions to $n(n+2) \equiv 0 \pmod{d}$, then

$$v(d) = \prod_{p \mid d} v(p).$$

Since $v(2) = 1$ while $v(p) = 2$ for primes $p > 2$, we deduce that $v(d) = 2^{\omega(d')}$, where d' denotes the largest odd divisor of d.

Thus, A_d counts the number of integers $n \leq x$ that belong to a certain $2^{\omega(d')}$ residue classes modulo d. Using the result of Problem 7.70, it follows that

$$\left| A_d - x \frac{2^{\omega(d')}}{d} \right| \leq 2^{\omega(d')}, \quad \text{so that in particular} \quad A_d = x \frac{2^{\omega(d')}}{d} + O(2^{\omega(d')}).$$

Using this in the second formula for $\pi_2(x, y)$ from Problem 7.71,

$$\pi_2(x, y) = \sum_{d \mid P} \mu(d) A_d$$

$$= x \sum_{d \mid P} \frac{2^{\omega(d')} \mu(d)}{d} + O\left(\sum_{d \mid P} 2^{\omega(d')} \right).$$

The error here is $O(3^{\pi(y)})$, since

$$\sum_{d \mid P} 2^{\omega(d')} \leq \sum_{d \mid P} 2^{\omega(d)} = \prod_{p \leq y} (1 + 2) = 3^{\pi(y)}.$$

Here we obtained the final equality by thinking of expanding out the product $\prod_{p \leq y} (1 + 2)$. That same reasoning shows that

$$\sum_{d \mid P} \frac{2^{\omega(d')} \mu(d)}{d} = \prod_{p \leq y} \left(1 - \frac{2^{\omega(p')}}{p} \right)$$

$$= \frac{1}{2} \prod_{2 < p \leq y} \left(1 - \frac{2}{p} \right).$$

Collecting our estimates, we have shown that for every $x, y \geq 2$,

$$\pi_2(x, y) = \frac{1}{2} x \prod_{2 < p \leq y} \left(1 - \frac{2}{p} \right) + O(3^{\pi(y)}). \tag{7.20}$$

What sort of bound for $\pi_2(x)$ can we extract from (7.20)? If p and $p+2$ are both prime, then either $p \leq y$, or p and $p+2$ have no prime factors up to y. Hence,

$$\pi_2(x) \leq \pi_2(x, y) + y$$

for every choice of $x, y \geq 2$. We can bound the main term on the right of (7.20) by observing that

$$\frac{1}{2}x \prod_{2 < p \leq y} \left(1 - \frac{2}{p}\right) \leq \frac{1}{2}x \prod_{2 < p \leq y} \left(1 - \frac{1}{p}\right)^2 = 2x \prod_{p \leq y} \left(1 - \frac{1}{p}\right)^2,$$

which is $O(x/(\log y)^2)$ by Problem 6.58. Thus,

$$\pi_2(x) = O\left(\frac{x}{(\log y)^2} + 3^{\pi(y)} + y\right).$$

Now take $y = \frac{1}{2}\log x$. Then $x/(\log y)^2 = O(x/(\log\log x)^2)$, as $x \to \infty$. Also, $3^{\pi(y)} \leq 3^y = x^{\frac{1}{2}\log 3} \leq x^{0.6}$, so that $3^{\pi(y)} = O(x/(\log\log x)^2)$. And certainly $y = O(x/(\log\log x)^2)$ as well. Therefore,

$$\pi_2(x) = O(x/(\log\log x)^2),$$

as $x \to \infty$.

Remark. This bound is not very interesting! In Step #8, we will prove that $\pi(x) = O(x/\log x)$, so that the estimate $\pi_2(x) = O(x/(\log\log x)^2)$ is worse than the trivial bound $\pi_2(x) \leq \pi(x)$! What prevents us from doing better is the $O(3^{\pi(y)})$ error term in our estimate for $\pi_2(x, y)$, which overwhelms the main term of size $x/(\log y)^2$ unless y is quite small.[1]

7.73

(a) We reason as in Problem 7.71. However, instead of the Principle of Inclusion-Exclusion, we apply the result of Problem 3.26, according to which we obtain an overestimate by stopping after an inclusion. The upshot is that for every even $k \geq 0$,

$$\pi_2(x, y) \leq A_1 - \sum_{p_1 \leq y} A_{p_1} + \cdots + \sum_{p_1 < \cdots < p_k \leq y} A_{p_1 \cdots p_k}$$

$$= \sum_{\substack{d | P \\ \omega(d) \leq k}} \mu(d) A_d.$$

---

[1] Up to now we have been crudely estimating $\pi(y)$ from above by y in these arguments. Of course, $\pi(y)$ is actually somewhat smaller than y—unfortunately, this turns out not to allow significantly larger choices of y above. Even taking advantage of the true size of $\pi(y)$, one cannot deduce anything better than $\pi_2(x) = O(x/(\log\log x)^2)$ by this method.

(b) In our solution to Problem 7.72, we found that

$$A_d = x \frac{2^{\omega(d')}}{d} + \theta_d 2^{\omega(d')}$$

where the real numbers θ_d satisfy $|\theta_d| \leq 1$. Inserting this into (a),

$$\pi_2(x, y) \leq \sum_{\substack{d|P \\ \omega(d) \leq k}} \left(x \frac{2^{\omega(d')} \mu(d)}{d} + \mu(d) \theta_d 2^{\omega(d')} \right)$$

$$\leq x \sum_{\substack{d|P \\ \omega(d) \leq k}} \frac{2^{\omega(d')} \mu(d)}{d} + E_2.$$

Also,

$$\sum_{\substack{d|P \\ \omega(d) \leq k}} \frac{2^{\omega(d')} \mu(d)}{d} = \sum_{d|P} \frac{2^{\omega(d')} \mu(d)}{d} - \sum_{\substack{d|P \\ \omega(d) > k}} \frac{2^{\omega(d')} \mu(d)}{d}$$

$$\leq \sum_{d|P} \frac{2^{\omega(d')} \mu(d)}{d} + E_1.$$

Collecting estimates completes the proof.

7.74 Inserting the definitions of $\zeta(2i)$ and $\zeta(2k - 2i)$,

$$\sum_{i=1}^{k-1} \zeta(2i)\zeta(2k - 2i) = \sum_{i=1}^{k-1} \sum_{m,n \geq 1} \frac{1}{m^{2i} n^{2k-2i}}$$

$$= \sum_{m,n \geq 1} \frac{1}{n^{2k}} \sum_{i=1}^{k-1} \left(\frac{n}{m} \right)^{2i}.$$

(Remember that there is never any danger in reordering series of positive terms.) Separating out the contribution of the terms where $m = n$,

$$\sum_{m,n \geq 1} \frac{1}{n^{2k}} \sum_{i=1}^{k-1} \left(\frac{n}{m} \right)^{2i} = (k - 1)\zeta(2k) + \sum_{\substack{m,n \geq 1 \\ m \neq n}} \frac{1}{n^{2k}} \sum_{i=1}^{k-1} \left(\frac{n}{m} \right)^{2i}.$$

The inner sum is a finite geometric series, and

$$\sum_{\substack{m,n\geq 1 \\ m\neq n}} \frac{1}{n^{2k}} \sum_{i=1}^{k-1} \left(\frac{n}{m}\right)^{2i} = \sum_{\substack{m,n\geq 1 \\ m\neq n}} \frac{1}{n^{2k}} \frac{(n/m)^{2k} - (n/m)^2}{(n/m)^2 - 1}$$

$$= \sum_{\substack{m,n\geq 1 \\ m\neq n}} \left(\frac{1}{m^{2k-2}} \frac{1}{n^2 - m^2} - \frac{1}{n^{2k-2}} \frac{1}{n^2 - m^2}\right)$$

$$= \sum_{\substack{m,n\geq 1 \\ m\neq n}} \left(\frac{1}{m^{2k-2}} \frac{1}{n^2 - m^2} + \frac{1}{n^{2k-2}} \frac{1}{m^2 - n^2}\right).$$

7.75 Let m be an integer with $1 \leq m \leq 2n$. Integrating by parts,

$$\int_0^1 u^m (1-u)^{2n-m}\, du = \int_0^1 u^m\, d\left(-\frac{1}{2n-(m-1)}(1-u)^{2n-(m-1)}\right)$$

$$= \frac{m}{2n-(m-1)} \int_0^1 u^{m-1}(1-u)^{2n-(m-1)}\, du.$$

Applying this with $m = n, n-1, n-2, \ldots, 1$, successively, we derive that

$$\int_0^1 u^n (1-u)^n\, du = \frac{n \cdot (n-1) \cdots 1}{(n+1) \cdot (n+2) \cdots 2n} \int_0^1 (1-u)^{2n}\, du$$

$$= \frac{1}{2n+1}\binom{2n}{n}^{-1}.$$

Since $\int_0^1 u^n (1-u)^n\, du \in \frac{1}{\mathrm{LCM}(2n+1)}\mathbb{Z}^+$, we may now deduce that

$$\mathrm{LCM}(2n+1) \geq \binom{2n}{n}(2n+1).$$

Let $R(n) = \frac{\binom{2n}{n}(2n+1)}{4^n}$. We find after a straightforward calculation that

$$\frac{R(n+1)}{R(n)} = \left(1 + \frac{1}{2n+2}\right).$$

Noting that $R(1) = \frac{3}{2}$, it follows that for all positive integers n,

$$R(n) = \frac{3}{2} \prod_{m=1}^{n-1} \left(1 + \frac{1}{2m+2}\right).$$

It is not hard to prove that the product on the right tends to infinity, as $n \to \infty$. For our purposes, it is enough to note that, by a direct calculation, the above right-hand side exceeds 4 for $n = 12$, and hence also for all $n \geq 12$. Thus, for $n \geq 12$,

$$\text{LCM}(2n + 1) \geq \binom{2n}{n}(2n + 1) > 4^{n+1} > 2^{2n+1},$$

while

$$\text{LCM}(2n + 2) \geq \text{LCM}(2n + 1) > 4^{n+1} = 2^{2n+2}.$$

We conclude that $\text{LCM}(n) > 2^n$ for all $n \geq 25$. (By computing the values of $\text{LCM}(n)$ for $n < 25$, it can be shown that in fact $\text{LCM}(n) > 2^n$ holds already once $n > 6$.)

Solutions to Step #8

8.76 Let p be an odd prime, and suppose for a contradiction that $r_{i+1} - r_i \geq p^{\frac{1}{2}+\epsilon}$ for some $i < \frac{p-1}{2}$. Then with N denoting the greatest integer smaller than $p^{\frac{1}{2}+\epsilon}$, all of $r_i + 1, \ldots, r_i + N$ are nonsquares modulo p. Thus,

$$\left| \sum_{n=r_i+1}^{r_i+N} \left(\frac{n}{p} \right) \right| = N.$$

However, by the Pólya–Vinogradov inequality (Problem 6.55),

$$\left| \sum_{n=r_i+1}^{r_i+N} \left(\frac{n}{p} \right) \right| = \left| \sum_{n=1}^{r_i+N} \left(\frac{n}{p} \right) - \sum_{n=1}^{r_i} \left(\frac{n}{p} \right) \right|$$

$$\leq \left| \sum_{n=1}^{r_i+N} \left(\frac{n}{p} \right) \right| + \left| \sum_{n=1}^{r_i} \left(\frac{n}{p} \right) \right| < 2\sqrt{p} \log p.$$

Thus, $N < 2\sqrt{p} \log p$. This contradicts the definition of N, once p is large enough in terms of ϵ. The proof that each $n_{i+1} - n_i < p^{\frac{1}{2}+\epsilon}$ is entirely analogous.

8.77 The first inequality is easy: $\prod_{p \leq x} p$ is a product of $\pi(x)$ distinct positive integers, and every such product is at least $\pi(x)!$. The second is equally clear, since $\prod_{p \leq x} p$ is a divisor of $\text{LCM}(x)$.

8.78 From Problem 7.68, $\text{LCM}(x) = \text{LCM}(\lfloor x \rfloor) \leq 4^{\lfloor x \rfloor} \leq 4^x$ for every real number $x \geq 1$. So from Problem 8.77,

$$\pi(x)! \leq 4^x, \quad \text{whence} \quad \log \pi(x)! \leq x \log 4. \tag{8.21}$$

© Springer Nature Switzerland AG 2021
P. Pollack, A. Singha Roy, *Steps into Analytic Number Theory*, Problem Books in
Mathematics, https://doi.org/10.1007/978-3-030-65077-3_23

We now explain how to deduce the desired upper bound on $\pi(x)$ from the estimate

$$\log m! = m \log m - m + O(\log(em)).$$

(This was shown back in Problem 2.12.) Fix a positive real number $\delta < 1$, and note that the result just recalled implies that all sufficiently large m,

$$\log m! > \delta m \log m.$$

Suppose that $\pi(x) > (\log 4 + \epsilon)\frac{x}{\log x}$. We will see that this leads to a contradiction for large x, provided δ was chosen sufficiently close to 1 to start with. Indeed, for all sufficiently all large x,

$$\log \pi(x) > \log x - \log \log x + \log \log 4 > \delta \log x,$$

and so

$$\log \pi(x)! > \delta \pi(x) \log \pi(x) > \delta \left((\log 4 + \epsilon)\frac{x}{\log x} \right) \cdot \delta \log x$$

$$= \delta^2 (\log 4 + \epsilon)x.$$

This contradicts (8.21) if $\delta^2 > \frac{\log 4}{\log 4 + \epsilon}$.

8.79 A prime p divides $\mathrm{LCM}(x)$ precisely when $p \le x$. For each such p, the exponent on p in the factorization of $\mathrm{LCM}(x)$ is the largest integer k with $p^k \le x$. That k is nothing other than $\lfloor \log x / \log p \rfloor$. Thus,

$$\mathrm{LCM}(x) = \prod_{p \le x} p^{\lfloor \log x / \log p \rfloor}, \quad \text{and} \quad \log \mathrm{LCM}(x) = \sum_{p \le x} \log p \left\lfloor \frac{\log x}{\log p} \right\rfloor.$$

Clearly,

$$\sum_{p \le x} \log p \left\lfloor \frac{\log x}{\log p} \right\rfloor \le \sum_{p \le x} \log p \cdot \frac{\log x}{\log p} = \sum_{p \le x} \log x = \pi(x) \log x.$$

8.80 Let n be the largest integer with $2n + 1 \le x$. For $x \ge 3$, we have $n \ge 1$, so that by Problems 7.69 and 8.79,

$$n \log 4 \le \log \mathrm{LCM}(2n + 1) \le \log \mathrm{LCM}(x) \le \pi(x) \log x.$$

But $n = \lfloor \frac{1}{2}(x - 1) \rfloor > \frac{1}{2}(x - 3)$, and so $n > \frac{1}{2}(1 - \epsilon)x$ for all large enough x. Thus, for all large x,

$$\pi(x) \geq \log 4 \cdot \frac{n}{\log x} > \log 2 \cdot (1-\epsilon)\frac{x}{\log x} > (\log 2 - \epsilon)\frac{x}{\log x}.$$

8.81 Suppose $K > 2$. For every fixed $\epsilon > 0$ and all large enough x,

$$\frac{\pi(Kx)}{\pi(x)} > \frac{(\log 2 - \epsilon) \cdot Kx/\log(Kx)}{(\log 4 + \epsilon) \cdot x/\log x}$$

$$= K\frac{\log 2 - \epsilon}{\log 4 + \epsilon} \cdot \frac{\log x}{\log Kx}. \qquad (8.22)$$

Since $K > 2$ and $\frac{\log 2}{\log 4} = \frac{1}{2}$, we can fix $\epsilon > 0$ small enough to ensure that $K\frac{\log 2 - \epsilon}{\log 4 + \epsilon} > 1$. Since the ratio $\frac{\log x}{\log Kx}$ tends to 1, (8.22) is then larger than 1 for all large x. In other words, $\pi(Kx) > \pi(x)$, and so the interval $(x, Kx]$ contains at least one prime.

8.82 We have already said that the highest power of p dividing $\mathrm{LCM}(x)$ is p^k, with k the largest integer satisfying $p^k \leq x$. Musing on this for a moment, we see that $\mathrm{LCM}(x)$ admits the factorization

$$\mathrm{LCM}(x) = \prod_{p^k \leq x} p,$$

where the product runs over all the numbers p^k with p prime, $k \geq 1$, and $p^k \leq x$. It follows that for $x \geq 1$,

$$\log\left(\prod_{m \leq x} \mathrm{LCM}(x/m)\right) = \log\left(\prod_{m \leq x}\prod_{p^k \leq x/m} p\right) = \sum_{m \leq x}\sum_{p^k \leq x/m} \log p$$

$$= \sum_{p^k \leq x} \log p \sum_{m \leq x/p^k} 1 = \sum_{p^k \leq x} \log p \left\lfloor \frac{x}{p^k} \right\rfloor.$$

This last expression can be rewritten as

$$\sum_{p \leq x} \log p \left(\left\lfloor \frac{x}{p} \right\rfloor + \left\lfloor \frac{x}{p^2} \right\rfloor + \cdots\right),$$

which we recognize as $\log\lfloor x\rfloor!$ (see the end of the solution to Problem 7.66). Now exponentiate.

8.83

(a) We begin by showing that χ is multiplicative. In fact, χ has the stronger property that

$$\chi(nm) = \chi(n)\chi(m) \qquad \text{for all pairs of integers } n, m. \qquad (8.23)$$

To see this, note that if 3 divides n or m, then both sides of (8.23) vanish. If both $n, m \equiv 1 \pmod 3$, or both $n, m \equiv -1 \pmod 3$, then both sides are 1. If $n \equiv 1 \pmod 3$ while $m \equiv -1 \pmod 3$, or vice-versa, both sides are -1.

Next, we notice that $\sum_{n=1}^{\infty} \frac{\chi(n)}{n^s}$ converges absolutely for real $s > 1$, since $\frac{|\chi(n)|}{n^s} \le \frac{1}{n^s}$ for all n.

These observations allow us to factor $\sum_{n=1}^{\infty} \frac{\chi(n)}{n^s}$, by the procedure used in the solution to Problem 3.22. (See the remark made at the end of the solution to Problem 3.25 for a general statement along these lines.) We obtain in this way that for all real $s > 1$,

$$\sum_{n=1}^{\infty} \frac{\chi(n)}{n^s} = \prod_p \left(1 + \frac{\chi(p)}{p^s} + \frac{\chi(p^2)}{p^{2s}} + \cdots \right). \qquad (8.24)$$

Because of (8.23), the factor corresponding to p in (8.24) is a geometric series with first term 1 and common ratio $\frac{\chi(p)}{p^s}$, and so has sum $\frac{1}{1 - \frac{\chi(p)}{p^s}}$.

(b) For each $s > 1$, the series for $L(s)$ is alternating with terms decreasing in absolute value. Thus,

$$L(s) = 1 - \left(\frac{1}{2^s} - \frac{1}{4^s} \right) - \left(\frac{1}{5^s} - \frac{1}{7^s} \right) + \cdots$$

$$\le 1 - 0 - 0 - 0 - \cdots = 1,$$

and

$$L(s) = \left(1 - \frac{1}{2^s} \right) + \left(\frac{1}{4^s} - \frac{1}{5^s} \right) + \left(\frac{1}{7^s} - \frac{1}{8^s} \right) + \cdots$$

$$\ge \left(1 - \frac{1}{2} \right) + 0 + 0 + 0 + \cdots = \frac{1}{2}.$$

That $\log L(s) = O(1)$ is now immediate: $\log \frac{1}{2} \le \log L(s) \le 0$.

(c) Taking logs,

$$\log L(s) = \log \prod_p \frac{1}{1 - \frac{\chi(p)}{p^s}} = \sum_p \log \frac{1}{1 - \frac{\chi(p)}{p^s}}.$$

Recalling that $\log \frac{1}{1-x} = x + \frac{x^2}{2} + \frac{x^3}{3} + \cdots$ when $|x| < 1$,

$$\sum_p \log \frac{1}{1 - \frac{\chi(p)}{p^s}} = \sum_p \sum_{k \ge 1} \frac{\chi(p^k)}{kp^{ks}}.$$

This double sum on p and k converges absolutely for each $s > 1$, since

$$\sum_p \sum_{k \geq 1} \frac{|\chi(p^k)|}{kp^{ks}} \leq \sum_p \sum_{k \geq 1} \frac{1}{kp^{ks}} = \sum_p \log \frac{1}{1 - \frac{1}{p^s}} = \log \zeta(s).$$

Thus, we may rearrange the terms to sum on k first. Doing so, we find that

$$\sum_p \sum_{k \geq 1} \frac{\chi(p^k)}{kp^{ks}} = \sum_p \frac{\chi(p)}{p^s} + \sum_{k \geq 2} \sum_p \frac{\chi(p^k)}{kp^{ks}},$$

where

$$\left| \sum_{k \geq 2} \sum_p \frac{\chi(p^k)}{kp^{ks}} \right| \leq \sum_{k \geq 2} \sum_p \frac{1}{kp^{ks}} < 1.$$

(See the solution to Problem 3.23 for the final inequality.) Hence,

$$\log L(s) = \sum_p \frac{\chi(p)}{p^s} + O(1)$$

$$= \sum_{p \equiv 1 \ (\mathrm{mod}\ 3)} \frac{1}{p^s} - \sum_{p \equiv -1 \ (\mathrm{mod}\ 3)} \frac{1}{p^s} + O(1).$$

From (b), $\log L(s) = O(1)$. Therefore,

$$\sum_{p \equiv 1 \ (\mathrm{mod}\ 3)} \frac{1}{p^s} - \sum_{p \equiv -1 \ (\mathrm{mod}\ 3)} \frac{1}{p^s} = O(1).$$

(d) For $1 < s < 2$, we have from Problem 3.24 that

$$\sum_{p \equiv 1 \ (\mathrm{mod}\ 3)} \frac{1}{p^s} + \sum_{p \equiv -1 \ (\mathrm{mod}\ 3)} \frac{1}{p^s} = \left(\sum_p \frac{1}{p^s} \right) - \frac{1}{3^s}$$

$$= \log \frac{1}{s - 1} + O(1).$$

Adding or subtracting the result of (c) from this, we obtain that for either choice of sign,

$$\sum_{p \equiv \pm 1 \ (\mathrm{mod}\ 3)} \frac{1}{p^s} = \frac{1}{2} \log \frac{1}{s - 1} + O(1).$$

8.84 Letting $s \downarrow 1$ in Problem 8.83(d), we see that $\sum_{p \equiv \pm 1 \ (\mathrm{mod} \ 3)} \frac{1}{p^s} \to \infty$, for either choice of sign. Hence, $\sum_{p \equiv \pm 1 \ (\mathrm{mod} \ 3)} \frac{1}{p}$ diverges.

To get the stated estimates for $\sum_{p \leq x, \ p \equiv \pm 1 \ (\mathrm{mod} \ 3)} \frac{1}{p}$, we adapt the solutions of Problems 4.34 and 4.35. First, let $\epsilon = \frac{1}{\log x \cdot \log \log x}$. Then for large x,

$$\sum_{\substack{p \leq x \\ p \equiv \pm 1 \ (\mathrm{mod} \ 3)}} \frac{1}{p} \leq \sum_{\substack{p \leq x \\ p \equiv \pm 1 \ (\mathrm{mod} \ 3)}} \frac{1}{p} \left(\frac{x}{p}\right)^{\epsilon} = x^{\epsilon} \sum_{p \equiv \pm 1 \ (\mathrm{mod} \ 3)} \frac{1}{p^{1+\epsilon}}$$

$$= \frac{1}{2} x^{\epsilon} \log \frac{1}{\epsilon} + O(x^{\epsilon}) = \frac{1}{2} \log \log x + \frac{1}{2} \log \log \log x + O(1).$$

On the other hand, with $\epsilon = \frac{\log \log x}{\log x}$,

$$\sum_{\substack{p \leq x \\ p \equiv \pm 1 \ (\mathrm{mod} \ 3)}} \frac{1}{p} \geq \sum_{\substack{p \leq x \\ p \equiv \pm 1 \ (\mathrm{mod} \ 3)}} \frac{1}{p^{1+\epsilon}} > \sum_{p \equiv \pm 1 \ (\mathrm{mod} \ 3)} \frac{1}{p^{1+\epsilon}} - \sum_{n > x} \frac{1}{n^{1+\epsilon}}$$

$$\geq \frac{1}{2} \log \frac{1}{\epsilon} - \frac{x^{-\epsilon}}{\epsilon} + O(1) = \frac{1}{2} \log \log x - \frac{1}{2} \log \log \log x + O(1).$$

8.85 Almost everything follows at once when we pick the correct χ. This time, we define $\chi : \mathbb{Z} \to \{-1, 0, 1\}$ by

$$\chi(n) = \begin{cases} 1 & \text{if } n \equiv 1 \ (\mathrm{mod} \ 4), \\ -1 & \text{if } n \equiv -1 \ (\mathrm{mod} \ 4), \\ 0 & \text{if } 2 \mid n. \end{cases}$$

We let

$$L(s) = \sum_{n=1}^{\infty} \frac{\chi(n)}{n^s}$$

$$= 1 - \frac{1}{3^s} + \frac{1}{5^s} - \frac{1}{7^s} + \cdots.$$

Then the assertion of Problem 8.83(a) holds as stated, with the same proof. Turning to (b), our previous proof now shows that $1 \geq L(s) \geq \frac{2}{3}$ for all $s > 1$; again, this implies $\log L(s) = O(1)$. Parts (c) and (d) hold, by essentially the same proofs as before, once "3" is changed to "4". With these assertions in hand, the proofs in Problem 8.84 go through essentially unchanged.

8.86 The sum expression for S_ℓ follows quickly from Problem 4.39. Interchanging the order of summation, we find that

$$S_\ell = \frac{1}{p} \sum_{k \bmod p} e^{-2\pi i k/p} \sum_{x_1,\ldots,x_\ell \bmod p} e^{2\pi i k(x_1^2+\cdots+x_\ell^2)/p}$$

$$= \frac{1}{p}\left(p^\ell + \sum_{\substack{k \bmod p \\ k \neq 0}} e^{-2\pi i k/p} \left(\sum_{x \bmod p} e^{2\pi i k x^2/p} \right)^\ell \right)$$

$$= \frac{1}{p}\left(p^\ell + \sum_{\substack{k \bmod p \\ k \neq 0}} e^{-2\pi i k/p} \left(\left(\frac{k}{p}\right) G \right)^\ell \right),$$

where we used the result of Problem 5.43 for the final step. To continue, we take cases according to the parity of ℓ. Suppose first that ℓ is odd. Then

$$\sum_{\substack{k \bmod p \\ k \neq 0}} e^{-2\pi i k/p} \left(\left(\frac{k}{p}\right) G \right)^\ell = G^\ell \sum_{k \bmod p} \left(\frac{k}{p}\right) e^{-2\pi i k/p}$$

$$= G^\ell \sum_{k \bmod p} \left(\frac{-k}{p}\right) e^{2\pi i k/p} = G^\ell \left(\frac{-1}{p}\right) \sum_{k \bmod p} \left(\frac{k}{p}\right) e^{2\pi i k/p}.$$

The number of square roots of k modulo p is given by $1 + \left(\frac{k}{p}\right)$. Hence, viewing $a^2 \equiv k$,

$$G = \sum_{a \bmod p} e^{2\pi i a^2/p} = \sum_{k \bmod p} \left(1 + \left(\frac{k}{p}\right)\right) e^{2\pi i k/p} = \sum_{k \bmod p} \left(\frac{k}{p}\right) e^{2\pi i k/p}.$$

$$(8.25)$$

Putting this back into the last display, and recalling that $G^2 = \left(\frac{-1}{p}\right)p$, we deduce that (for ℓ odd)

$$S_\ell = p^{\ell-1} + p^{-1} G^{\ell+1}\left(\frac{-1}{p}\right)$$

$$= p^{\ell-1} + G^{\ell-1}$$

$$= p^{\ell-1} + \left(\left(\frac{-1}{p}\right)p\right)^{\frac{\ell-1}{2}}$$

$$= p^{\ell-1} + (-1)^{\frac{p-1}{2} \cdot \frac{\ell-1}{2}} p^{\frac{\ell-1}{2}}.$$

$$(8.26)$$

(We used here that $\left(\frac{-1}{p}\right) = (-1)^{\frac{p-1}{2}}$.) If instead ℓ is even, then

$$\sum_{\substack{k \bmod p \\ k \not\equiv 0}} e^{-2\pi i k / p} \left(\left(\frac{k}{p} \right) G \right)^{\ell} = G^{\ell} \sum_{\substack{k \bmod p \\ k \not\equiv 0}} e^{-2\pi i k / p} = -G^{\ell} = -\left(\left(\frac{-1}{p} \right) p \right)^{\ell/2},$$

and so

$$S_{\ell} = p^{\ell-1} - \left(\frac{-1}{p} \right)^{\ell/2} p^{\frac{1}{2}\ell-1}.$$

Remarks.

(a) The formula (8.26) for S_{ℓ} (ℓ odd) yields a slick proof of the famous Law of Quadratic Reciprocity.

Assume ℓ and p are odd primes, $\ell \neq p$. From (8.26) and Euler's criterion for quadratic residues,

$$S_{\ell} \equiv 1 + (-1)^{\frac{p-1}{2} \cdot \frac{\ell-1}{2}} \left(\frac{p}{\ell} \right) \pmod{\ell}.$$

Here is another way of computing S_{ℓ} (mod ℓ). Observe that if $(x_1, \ldots, x_{\ell})$ is a tuple counted in the definition of S_{ℓ}, so is its cyclic shift $(x_2, \ldots, x_{\ell-1}, x_1)$. Call two tuples equivalent if one can be obtained from the other by a finite number of cyclic shifts; it is easy to see that this is an equivalence relation, and that each equivalence class has size dividing ℓ, so is either 1 or ℓ. Therefore, S_{ℓ} is congruent, mod ℓ, to the number of classes of size 1. That number is just

$$\#\{x \in \mathbb{Z}_p : \ell x^2 = 1\} = \#\{\text{square roots of } 1/\ell \bmod p\}$$

$$= \#\{\text{square roots of } \ell \bmod p\} = 1 + \left(\frac{\ell}{p} \right).$$

Thus, modulo ℓ,

$$1 + \left(\frac{\ell}{p} \right) \equiv 1 + (-1)^{\frac{p-1}{2} \cdot \frac{\ell-1}{2}} \left(\frac{p}{\ell} \right).$$

Subtracting 1 and multiplying both sides by $\left(\frac{p}{\ell} \right)$, we obtain

$$\left(\frac{p}{\ell} \right)\left(\frac{\ell}{p} \right) \equiv (-1)^{\frac{p-1}{2} \cdot \frac{\ell-1}{2}}.$$

Since both sides here are ± 1, this congruence mod ℓ must be a genuine equality of integers, completing the proof of Quadratic Reciprocity.

(b) Problem 8.86 demonstrates how the seemingly trivial act of rewriting a quantity as an exponential sum can yield nontrivial results. Here is another entertaining example.

To set up, recall that every element of $\mathbb{Z}_p$ is a sum of two squares. Here is an easy proof: We assume p is odd, since the result is obvious when $p = 2$. Now take any $r \in \mathbb{Z}_p$. The sets $A = \{r - a^2 : a \in \mathbb{Z}_p\}$ and $B = \{b^2 : b \in \mathbb{Z}_p\}$ both have size $\frac{p+1}{2} > \frac{p}{2}$, so A and B cannot be disjoint. Hence, there are $a, b \in \mathbb{Z}_p$ with $r - a^2 = b^2$, and $r = a^2 + b^2$.

By contrast, it is not always the case that every element of $\mathbb{Z}_p$ is a sum of two *cubes*. For instance, when $p = 7$, the cubes mod p are -1, 0, and 1, so that 3 and 4 are not representable. We now show that 7 is the exception rather than the rule: For all sufficiently large primes p, every element of $\mathbb{Z}_p$ is a sum of two cubes. Our argument works as soon as $p \geq 233$; with more effort, one can show that $p > 7$ suffices.

If $p = 3$, every element of $\mathbb{Z}_p$ is a cube, the cube of itself. When $p \equiv 2 \pmod 3$, it is still the case that every $a \in \mathbb{Z}_p$ is a cube: $(a^{(2p-1)/3})^3 = a^{2p-1} = a^p a^{p-1} = a^p = a$. So the problem is only interesting when $p \equiv 1 \pmod 3$, which we now assume.

Let $a \in \mathbb{Z}_p$ with $a \neq 0$. (Clearly, 0 is a sum of two cubes.) We look at the solutions to $x^3 + y^3 - az^3 = 0$, with $x, y, z \in \mathbb{Z}_p$. Note that if there is a solution with $z \neq 0$, then $(x/z)^3 + (y/z)^3 = a$.

The number of solutions to $x^3 + y^3 - az^3 = 0$ having $z = 0$ is easily seen to be $1 + 3(p-1)$: There is the solution $x = y = 0$, and for each nonzero $x \in \mathbb{Z}_p$, there are 3 values of y with $x^3 + y^3 - a \cdot 0^3 = 0$ (namely, $y = -\omega x$, where $\omega \in \mathbb{Z}_p$ satisfies $\omega^3 = 1$). Thus, we are guaranteed a is a sum of two cubes as long as the count N of solutions to $x^3 + y^3 - az^3 = 0$ exceeds $1 + 3(p - 1)$. We write N as an exponential sum:

$$ N = \frac{1}{p} \sum_{x,y,z \bmod p} \sum_{k \bmod p} e^{2\pi i k(x^3 + y^3 - az^3)/p}. $$

Now setting

$$ \tilde{G}_k := \sum_{a \bmod p} e^{2\pi i k a^3/p} $$

(these are "cubic Gauss sums"), we see that

$$ N = \frac{1}{p} \sum_{k \bmod p} \tilde{G}_k^2 \cdot \tilde{G}_{-ak} = p^2 + \frac{1}{p} \sum_{\substack{k \bmod p \\ k \not\equiv 0}} \tilde{G}_k^2 \cdot \tilde{G}_{-ak}. $$

We show below that $|\tilde{G}_k| \leq \sqrt{6p}$ for all $k \not\equiv 0 \pmod p$. It follows immediately that

$$ N \geq p^2 - \frac{1}{p}(p - 1)(\sqrt{6p})^3 $$

$$ > p^2 - (\sqrt{6})^3 p^{3/2} > p^2 - 15 p^{3/2}. $$

The last expression exceeds $1 + 3(p - 1)$ for all $p \geq 233$. So for $p \geq 233$, every $a \in \mathbb{Z}_p$ is a sum of two cubes.

To prove the claimed inequality for $|\tilde{G}_k|$, we observe that

$$ \sum_{\substack{k \bmod p \\ k \not\equiv 0}} |\tilde{G}_k|^2 = \left(\sum_{k \bmod p} |\tilde{G}_k|^2 \right) - p^2, $$

and that

$$ \sum_{k \bmod p} |\tilde{G}_k|^2 = \sum_{k \bmod p} \tilde{G}_k \cdot \overline{\tilde{G}_k} $$

$$ = \sum_{k \bmod p} \sum_{x,y \bmod p} e^{2\pi i k(x^3 - y^3)/p} $$

$$= \sum_{x,y \bmod p} \sum_{k \bmod p} e^{2\pi i k(x^3 - y^3)/p}$$

$$= p \cdot \#\{(x, y) \in (\mathbb{Z}_p)^2 : x^3 = y^3\}$$

$$= p(1 + 3(p - 1)).$$

Thus,

$$\sum_{\substack{k \bmod p \\ k \not\equiv 0}} |\tilde{G}_k|^2 = p(1 + 3(p - 1)) - p^2 = 2p(p - 1).$$

Now the value of $\tilde{G}_k$ depends only on the coset of k with respect to the subgroup of cubes in $\mathbb{U}_p$. So using a, b, c for representatives of the three cosets,

$$(|\tilde{G}_a|^2 + |\tilde{G}_b|^2 + |\tilde{G}_c|^2) \cdot \frac{p - 1}{3} = 2p(p - 1),$$

and so

$$|\tilde{G}_a|^2 + |\tilde{G}_b|^2 + |\tilde{G}_c|^2 = 6p.$$

It is now immediate that $|\tilde{G}_a|, |\tilde{G}_b|, |\tilde{G}_c| \leq \sqrt{6p}$, finishing the proof of the claim.

It is not difficult to modify the above arguments above to prove that for each fixed d, and all primes $p > p_0(d)$, every element of $\mathbb{Z}_p$ is a sum of two dth powers.

For an extensive discussion of how Gauss sums (and closely allied concepts) can be used to count solutions to equations over finite fields, see Chapter II of [2] or Chapter 8 of [1].

References

1. Kenneth Ireland and Michael A. Rosen, *A classical introduction to modern number theory.* Second edition. Graduate Texts in Mathematics, vol. 84. Springer-Verlag, New York, 1990.
2. Wolfgang M. Schmidt, *Equations over finite fields. An elementary approach.* Lecture Notes in Mathematics, vol. 536. Springer-Verlag, Berlin-New York, 1976.

Solutions to Step #9

9.87 Let $n \in \mathbb{Z}^+$, and factor $n = \prod_p p^{e_p}$. A given $d \in \mathbb{Z}^+$ has d^2 dividing n if and only the exponent of p in d is at most $e_p/2$, for each p, i.e., if and only if d divides $n' := \prod_{p \mid n} p^{\lfloor e_p/2 \rfloor}$. Therefore,

$$\sum_{d^2 \mid n} \mu(d) = \sum_{d \mid n'} \mu(d) = \begin{cases} 1 & \text{if } n' = 1, \\ 0 & \text{otherwise.} \end{cases}$$

Now $n' = 1$ if and only if n is squarefree. Hence, $\sum_{d^2 \mid n} \mu(d)$ is the indicator function of the squarefree numbers—but that description also fits $|\mu(n)|$.

9.88 From the expression for $|\mu(n)|$ in the last problem,

$$\sum_{n \le x} |\mu(n)| = \sum_{n \le x} \sum_{d^2 \mid n} \mu(d) = \sum_{d \le \sqrt{x}} \mu(d) \sum_{\substack{n \le x \\ d^2 \mid n}} 1 = \sum_{d \le \sqrt{x}} \mu(d) \left\lfloor \frac{x}{d^2} \right\rfloor.$$

Continuing,

$$\sum_{d \le \sqrt{x}} \mu(d) \left\lfloor \frac{x}{d^2} \right\rfloor = \sum_{d \le \sqrt{x}} \mu(d) \left(\frac{x}{d^2} + O(1) \right)$$

$$= x \sum_{d=1}^{\infty} \frac{\mu(d)}{d^2} - x \sum_{d > \sqrt{x}} \frac{\mu(d)}{d^2} + O(\sqrt{x}).$$

By Exercise 2.11, $\left| x \sum_{d > \sqrt{x}} \frac{\mu(d)}{d^2} \right| \le x \sum_{d > \sqrt{x}} \frac{1}{d^2} = O(\sqrt{x})$. And from the solution to Exercise 3.25, $\sum_{d=1}^{\infty} \frac{\mu(d)}{d^2} = \zeta(2)^{-1}$. Thus,

© Springer Nature Switzerland AG 2021
P. Pollack, A. Singha Roy, *Steps into Analytic Number Theory*, Problem Books in Mathematics, https://doi.org/10.1007/978-3-030-65077-3_24

$$\sum_{n \leq x} |\mu(n)| = \frac{1}{\zeta(2)} x + O(\sqrt{x}).$$

The probability claim, interpreted as a statement about the limiting probability of a randomly chosen number from $[1, N]$ being squarefree (as $N \to \infty$), follows immediately.

All of this is easily generalized to kth-power-free numbers, for any fixed $k \geq 2$. In this case, the characteristic function can be written as $\sum_{d^k|n} \mu(d)$. Following the above arguments shows that the count of kth-power-free numbers in $[1, x]$ is

$$\frac{x}{\zeta(k)} + O(x^{1/k}),$$

so that the probability of being kth-power-free is $1/\zeta(k)$.

9.89 Since $\omega(d') \leq \omega(d)$, the first inequality is clear. That

$$\sum_{\substack{d|P \\ \omega(d)>k}} \frac{2^{\omega(d)}}{d} \leq \sum_{d|P} \frac{2^{\omega(d)}}{d} e^{\omega(d)-k}$$

is also easy: Each d contributing to the left sum contributes at least as much to the sum on the right, since $e^{\omega(d)-k} > e^0$ for these d. And those d contributing to the right-hand sum but not the left make a nonnegative contribution, so are harmless for the desired inequality. Continuing on, the factorization

$$\sum_{d|P} \frac{2^{\omega(d)}}{d} e^{\omega(d)-k} = e^{-k} \prod_{p \leq y} \left(1 + \frac{2e}{p}\right)$$

follows immediately upon expanding out the right-hand side. Thus, it remains only to prove that

$$\prod_{p \leq y} \left(1 + \frac{2e}{p}\right) \leq (\log y)^6$$

for large enough x. Since y is a function of x that tends to infinity as $x \to \infty$, it is enough to prove the inequality for large enough y. For this we use Problem 4.34 and the inequality $1 + t \leq e^t$: For large y,

$$\prod_{p \leq y} \left(1 + \frac{2e}{p}\right) \leq \exp\left(2e \sum_{p \leq y} \frac{1}{p}\right)$$

$$\leq \exp(2e(\log \log y + 2 \log \log \log y))$$

$$= (\log y)^{2e} (\log \log y)^{4e}$$

$$< (\log y)^6.$$

In the final line, it is important that $2e = 5.43 \cdots < 6$.

9.90 If d divides P, then d is a squarefree product of primes not exceeding y. Writing $d = p_1 \cdots p_r$ with $r = \omega(d)$, it is then immediate that $d \leq y^r$. Thus, $d \leq y^k$ when $\omega(d) \leq k$, and

$$E_2 = \sum_{\substack{d|P \\ \omega(d) \leq k}} 2^{\omega(d')} \leq \sum_{\substack{d|P \\ \omega(d) \leq k}} 2^{\omega(d)} \leq \sum_{d \leq y^k} 2^{\omega(d)} \leq y^k \max_{d \leq y^k} 2^{\omega(d)}.$$

(At the last step, we bounded the sum by the number of summands multiplied by the largest term.) Since $2^{\omega(d)} \leq \tau(d) \leq d$ (the final bound being trivial), $\max_{d \leq y^k} 2^{\omega(d)} \leq y^k$, finishing the proof.

9.91 Piecing together what we know so far,

$$\pi_2(x, y) \leq \frac{1}{2}x \prod_{2 < p \leq y} \left(1 - \frac{2}{p}\right) + xE_1 + E_2$$

$$= O\left(\frac{x}{(\log y)^2} + xe^{-k}(\log y)^6 + y^{2k}\right).$$

Here the product on p has been estimated as in the solution to Problem 7.72. For the specified choices of y and k, and all large enough x,

$$\frac{x}{(\log y)^2} = O\left(\frac{x}{(\log x)^2}(\log \log x)^2\right),$$

$$xe^{-k}(\log y)^6 \leq xe^{-9 \log \log x}(\log x)^6 = x/(\log x)^3,$$

and

$$y^{2k} \leq (x^{\frac{1}{100 \log \log x}})^{20 \log \log x} = x^{1/5}.$$

Among the right-hand expressions, $\frac{x}{(\log x)^2}(\log \log x)^2$ dominates, so that

$$\pi_2(x, x^{\frac{1}{100 \log \log x}}) = O\left(\frac{x}{(\log x)^2}(\log \log x)^2\right).$$

Since $\pi_2(x) \leq \pi_2(x, y) + y$ for all choices of $x, y \geq 2$, we deduce that also

$$\pi_2(x) = O\left(\frac{x}{(\log x)^2}(\log \log x)^2\right).$$

9.92 Suppose for a contradiction that $\sum_{p \in \mathcal{P}} \frac{1}{p}$ diverges. Then by Exercise 3.28, there are arbitrarily large values of x with $\pi_2(x) > x/(\log x)^{1.01}$. But this contradicts the upper bound on $\pi_2(x)$ shown in Problem 9.91.

Remark. The bound on $\pi_2(x)$ in Problem 9.91 was published by Brun in 1919. A year later, he succeeded in proving the somewhat sharper estimate $\pi_2(x) = O(x/(\log x)^2)$. It is widely believed that $\lim_{x \to \infty} \frac{\pi_2(x)}{x/(\log x)^2}$ exists and is positive (a special case of very-plausible seeming conjectures of Hardy–Littlewood), so that Brun's sharper upper bound is probably best possible. However, it remains a notorious open problem (the "twin prime conjecture") to even show that $\pi_2(x) \to \infty$ as $x \to \infty$.

9.93 Let a/b and a'/b' be reduced fractions with

$$0 < a, a', b, b' < \sqrt{p}, \tag{9.27}$$

and suppose that $a/b = a'/b'$ in $\mathbb{Z}_p$. Then $p \mid ab' - a'b$. But (9.27) implies that $|ab' - a'b| < p$, so it must be that $ab' - a'b = 0$, forcing $a/b = a'/b'$ in $\mathbb{Q}$. Since a/b and a'/b' are reduced, $a = a'$ and $b = b'$. Consequently, the reduced fractions a/b with $0 < a, b < \sqrt{p}$ represent distinct elements of $\mathbb{Z}_p$.

Counting these fractions a/b is equivalent to counting coprime lattice points in the box $(0, N] \times (0, N]$, where $N := \lfloor \sqrt{p} \rfloor$. As shown in the solution to Problem 4.38, this count is

$$\frac{1}{\zeta(2)} N^2 + O(N \log(eN)).$$

Since $\zeta(2) < 2$, this quantity exceeds $\frac{1}{2} p$ for large enough p, meaning that more than half the elements of $\mathbb{Z}_p$ are represented by some such fraction.

9.94 Suppose for a contradiction that every integer in $(0, \sqrt{p})$ reduces to a square modulo p. Then all of the fractions a/b, with $0 < a, b < \sqrt{p}$, represent nonzero squares in $\mathbb{Z}_p$. But the number of elements of $\mathbb{Z}_p$ represented by those fractions is $> \frac{1}{2} p$ for large p, whereas the number of nonzero squares in $\mathbb{Z}_p$ is $\frac{p-1}{2} < \frac{1}{2} p$.

Remark. By keeping careful track of the error terms, one can show by this method that the least positive nonsquare modulo p is smaller than $\sqrt{p}$ for every prime $p > 23$.

9.95 We recognize that $\frac{1}{n^2 - m^2} = \frac{1}{2m} \left(\frac{1}{n-m} - \frac{1}{n+m} \right)$ and rewrite

$$\sum_{\substack{n \in \mathbb{Z}^+ \\ n \neq m}} \frac{1}{n^2 - m^2} = \lim_{N \to \infty} \frac{1}{2m} \left(\sum_{\substack{1 \leq n \leq N \\ n \neq m}} \frac{1}{n - m} - \sum_{\substack{1 \leq n \leq N \\ n \neq m}} \frac{1}{n + m} \right).$$

When $N \geq m$,

$$\sum_{\substack{1 \le n \le N \\ n \ne m}} \frac{1}{n-m} - \sum_{\substack{1 \le n \le N \\ n \ne m}} \frac{1}{n+m} = -\left(\frac{1}{m-1} + \frac{1}{m-2} + \cdots + 1\right)$$

$$+ \left(1 + \frac{1}{2} + \cdots + \frac{1}{N-m}\right) - \left(\frac{1}{m+1} + \frac{1}{m+2} + \cdots + \frac{1}{m+N} - \frac{1}{2m}\right).$$

Now suppose that $N \ge 2m+1$. Then in the second displayed line, the fractions $\frac{1}{m+1}, \frac{1}{m+2}, \ldots, \frac{1}{N-m}$ appear with a $+$ sign in the first group of terms but with a $-$ sign in the second. Canceling, the above right-hand side becomes

$$-\left(\frac{1}{m-1} + \frac{1}{m-2} + \cdots + 1\right) + \left(1 + \frac{1}{2} + \cdots + \frac{1}{m}\right)$$

$$-\left(\frac{1}{N-m+1} + \frac{1}{N-m+2} + \cdots + \frac{1}{N+m} - \frac{1}{2m}\right).$$

In the first line of this new display, everything cancels except $\frac{1}{m}$. The subtraction of $-\frac{1}{2m}$ in the second line gives us an additional $\frac{1}{2m}$, while $\frac{1}{N-m+1}, \frac{1}{N-m+2}, \ldots, \frac{1}{N+m}$ make a contribution bounded in absolute value by $2m/(N-m+1)$. Therefore,

$$\frac{1}{2m} \left(\sum_{\substack{1 \le n \le N \\ n \ne m}} \frac{1}{n-m} - \sum_{\substack{1 \le n \le N \\ n \ne m}} \frac{1}{n+m} \right) = \frac{1}{2m} \left(\frac{1}{m} + \frac{1}{2m} + O\left(\frac{m}{N-m+1} \right) \right)$$

$$= \frac{3}{4m^2} + O\left(\frac{1}{N-m+1} \right).$$

Letting $N \to \infty$, we conclude that $\sum_{\substack{n \in \mathbb{Z}^+ \\ n \ne m}} \frac{1}{n^2 - m^2} = \frac{3}{4m^2}$.

9.96 The series converges if and only if $\lambda \ge 1$.

Take first the case when $\lambda = 1$. We reorder the terms according to the value of $P(n)$ (reordering is harmless as we deal with a series of positive terms):

$$\sum_{n>1} \frac{1}{n P(n)} = \sum_p \frac{1}{p} \sum_{n:\, P(n)=p} \frac{1}{n}$$

$$= \sum_p \frac{1}{p^2} \sum_{m=1 \text{ or } P(m) \le p} \frac{1}{m}$$

$$= \sum_p \frac{1}{p^2} \prod_{\ell \le p} \left(1 + \frac{1}{\ell} + \frac{1}{\ell^2} + \cdots\right),$$

where in this final product ℓ runs over the primes not exceeding p. For large enough p,

$$\prod_{\ell \le p} \left(1 + \frac{1}{\ell} + \frac{1}{\ell^2} + \dots\right) = \prod_{\ell \le p} \left(1 + \frac{1}{\ell - 1}\right) \le \prod_{\ell \le p} \left(1 + \frac{2}{\ell}\right)$$

$$\le \exp\left(\sum_{\ell \le p} \frac{2}{\ell}\right) \le \exp(2(\log \log p + 2 \log \log \log p)) < (\log p)^3.$$

The sum $\sum_p \frac{(\log p)^3}{p^2}$ certainly converges, since the larger sum $\sum_{n \ge 2} \frac{(\log n)^3}{n^2}$ converges. This establishes convergence of the initial series for $\lambda = 1$, and so also for $\lambda \ge 1$.

If $\lambda \le 0$, then the terms $\frac{1}{n^\lambda P(n)}$ do not tend to zero (consider n of the form 2^k), so the series diverges.

Finally, we suppose that $0 < \lambda < 1$. Proceeding as above,

$$\sum_{n>1} \frac{1}{n^\lambda P(n)} = \sum_p \frac{1}{p^{1+\lambda}} \prod_{\ell \le p} \left(1 + \frac{1}{\ell^\lambda} + \frac{1}{\ell^{2\lambda}} + \dots\right).$$

The series expansion $\log \frac{1}{1-t} = t + \frac{1}{2}t^2 + \dots$ tells us that

$$\log \prod_{\ell \le p} \left(1 + \frac{1}{\ell^\lambda} + \frac{1}{\ell^{2\lambda}} + \dots\right) = \sum_{\ell \le p} \log \frac{1}{1 - \frac{1}{\ell^\lambda}} \ge \sum_{\ell \le p} \frac{1}{\ell^\lambda} \ge \frac{1}{p^\lambda} \pi(p).$$

We know that $\pi(p) > \frac{1}{2} \frac{p}{\log p}$ for all large p (see Problem 8.80). Consequently, for all large primes p,

$$\frac{1}{p^\lambda} \pi(p) > \frac{1}{2} p^{1-\lambda} / \log p > 2 \log p.$$

For these p, the product $\prod_{\ell \le p} \left(1 + \frac{1}{\ell^\lambda} + \frac{1}{\ell^{2\lambda}} + \dots\right) > p^2$. It follows that the terms of the series

$$\sum_p \frac{1}{p^{1+\lambda}} \prod_{\ell \le p} \left(1 + \frac{1}{\ell^\lambda} + \frac{1}{\ell^{2\lambda}} + \dots\right)$$

tend to infinity (not zero!), and so we again have the desired divergence.

Solutions to Step #10

10.97

(a) By Legendre's formula,

$$\log\lfloor x\rfloor! = \log\prod_{p\le x} p^{\sum_{k\ge 1}\lfloor x/p^k\rfloor} = \sum_{p\le x}\left(\sum_{k\ge 1}\left\lfloor\frac{x}{p^k}\right\rfloor\right)\log p$$

$$= \sum_{p\le x}\left\lfloor\frac{x}{p}\right\rfloor\log p + \sum_{p\le x}\log p\sum_{k\ge 2}\left\lfloor\frac{x}{p^k}\right\rfloor.$$

Since

$$0 \le \sum_{p\le x}\log p\sum_{k\ge 2}\left\lfloor\frac{x}{p^k}\right\rfloor \le x\sum_{p\le x}\log p\sum_{k\ge 2}\frac{1}{p^k}$$

$$= x\sum_{p\le x}\frac{\log p}{p^2-p} \le x\sum_{n\ge 2}\frac{\log n}{n^2-n},$$

and $\sum_{n\ge 2}\frac{\log n}{n^2-n} < \infty$, it follows that

$$\log\lfloor x\rfloor! = x\sum_{p\le x}\frac{\log p}{p} - \sum_{p\le x}\log p\left(\frac{x}{p}-\left\lfloor\frac{x}{p}\right\rfloor\right) + O(x).$$

Finally,

$$0 \le \sum_{p\le x}\log p\left(\frac{x}{p}-\left\lfloor\frac{x}{p}\right\rfloor\right) \le \sum_{p\le x}\log p \le \log\mathrm{LCM}(x) \le x\log 4,$$

© Springer Nature Switzerland AG 2021
P. Pollack, A. Singha Roy, *Steps into Analytic Number Theory*, Problem Books in
Mathematics, https://doi.org/10.1007/978-3-030-65077-3_25

so that $\sum_{p\leq x} \log p (\frac{x}{p} - \lfloor \frac{x}{p} \rfloor) = O(x)$.

(b) Recall from Problem 2.12 that $\log \lfloor x \rfloor! = x \log x + O(x)$. (In fact, the estimate there was somewhat sharper.) Comparing with (a), $x \sum_{p\leq x} \frac{\log p}{p} = x \log x + O(x)$. Divide by x.

10.98 We apply the summation by parts formula established in Problem 6.60.

To set things up, we let $a_n = \frac{\log n}{n}$ when n is prime, and $a_n = 0$ otherwise. We let $f(t) = \frac{1}{\log t}$ for $t \geq 2$. We extend f to all of $[1, \infty)$ in such a way that $f'(t)$ is continuous for all $t \geq 1$. (The choice of extension is unimportant, as will become clear shortly.) Finally, we let $S(x) := \sum_{p\leq x} \frac{\log p}{p}$. Since $S(t)$ vanishes when $t < 2$, the summation by parts formula yields

$$\sum_{p\leq x} \frac{1}{p} = \sum_{n\leq x} a_n f(n) = S(x) f(x) - \int_1^x S(t) f'(t)\, dt$$

$$= S(x) f(x) - \int_2^x S(t) f'(t)\, dt = \frac{S(x)}{\log x} + \int_2^x \frac{S(t)}{t(\log t)^2}\, dt.$$

Write $S(t) = \log t + E(t)$, so that $E(t) = O(1)$ for all $t \geq 2$. Then, for $x \geq 2$,

$$\frac{S(x)}{\log x} = 1 + \frac{E(x)}{\log x} = 1 + O\left(\frac{1}{\log x}\right).$$

Also,

$$\int_2^x \frac{S(t)}{t(\log t)^2}\, dt = \int_2^x \left(\frac{1}{t \log t} + \frac{E(t)}{t(\log t)^2}\right) dt$$

$$= \log \log x - \log \log 2 + \int_2^x \frac{E(t)}{t(\log t)^2}\, dt.$$

Since $E(t) = O(1)$, the improper integral $\int_2^\infty \frac{E(t)}{t(\log t)^2}\, dt$ converges absolutely, and

$$\int_2^x \frac{E(t)}{t(\log t)^2}\, dt = \int_2^\infty \frac{E(t)}{t(\log t)^2}\, dt - \int_x^\infty \frac{E(t)}{t(\log t)^2}\, dt$$

$$= \int_2^\infty \frac{E(t)}{t(\log t)^2}\, dt + O\left(\int_x^\infty \frac{dt}{t(\log t)^2}\right)$$

$$= \int_2^\infty \frac{E(t)}{t(\log t)^2}\, dt + O\left(\frac{1}{\log x}\right).$$

Piecing the puzzle back together,

$$\sum_{p \le x} \frac{1}{p} = \left(1 + O\left(\frac{1}{\log x}\right)\right) +$$

$$\left(\log\log x - \log\log 2 + \int_2^\infty \frac{E(t)}{t(\log t)^2}\, dt + O\left(\frac{1}{\log x}\right)\right)$$

$$= \log\log x + C + O\left(\frac{1}{\log x}\right).$$

Remark. The constant C appearing here is usually referred to as the "Meissel–Mertens constant". It can be shown that

$$C = \gamma + \sum_p \left\{\log\left(1 - \frac{1}{p}\right) + \frac{1}{p}\right\}$$

where γ is the Euler–Mascheroni constant from Exercise 1.7. A consequence of this representation of C (or really, an equivalent form of that same result) is "Mertens' product theorem", which states that

$$\lim_{x \to \infty} e^\gamma \log x \prod_{p \le x} \left(1 - \frac{1}{p}\right) = 1.$$

Numerically, $C = 0.26149721\ldots$.

10.99

(a) A number $n \le x$ can have at most one prime factor $> \sqrt{x}$, for if it had two, say p and q, then we would have $n \ge pq > x$. So to count the $n \le x$ with a prime factor $> \sqrt{x}$, it suffices to add up, for each prime $p \in (\sqrt{x}, x]$, the number of multiples of $p \le x$. For large x, this is

$$\sum_{\sqrt{x} < p \le x} \left\lfloor \frac{x}{p} \right\rfloor = \sum_{\sqrt{x} < p \le x} \left(\frac{x}{p} + O(1)\right) = x \sum_{\sqrt{x} < p \le x} \frac{1}{p} + O(x/\log x),$$

using in the last step that the number of primes up to x is $O(x/\log x)$. By Problem 10.98,

$$\sum_{\sqrt{x} < p \le x} \frac{1}{p}$$

$$= \left(\log\log x + C + O\left(\frac{1}{\log x}\right)\right) - \left(\log\log \sqrt{x} + C + O\left(\frac{1}{\log \sqrt{x}}\right)\right)$$

$$= \log\log x - \log\log \sqrt{x} + O(1/\log x)$$

$$= \log 2 + O(1/\log x).$$

Hence, the number of $n \leq x$ having a prime factor $> \sqrt{x}$ is $x \log 2 + O(x/\log x)$. This means that the proportion of such n in $[1, x]$ tends to $\log 2$, which is $\approx 69.3\%$.

(b) We try $y = x^\alpha$ for an α to be decided upon. By (a), we should choose $\alpha > \frac{1}{2}$. For these α, the same reasoning as in (a) shows that the limiting proportion of $n \in [1, x]$ with a prime factor $> x^\alpha$ is

$$\log \log x - \log \log x^\alpha = \log \frac{1}{\alpha}.$$

Setting $\log \frac{1}{\alpha} = \frac{1}{2}$ and solving, $\alpha = e^{-1/2}$. So $y = x^{1/\sqrt{e}}$ works.

Remark. Positive integers n with all of their prime factors $\leq y$ are commonly referred to as "y-smooth numbers", and the count of such $n \leq x$ is usually written $\Psi(x, y)$. That is,

$$\Psi(x, y) := \#\{n \leq x : p \mid n \Rightarrow p \leq y\}.$$

Understanding the distribution of smooth numbers is crucial for many problems in analytic and computational number theory. For example, results on $\Psi(x, y)$ are needed in the analysis of the running times of the fastest-known integer factorization algorithms.

Our solution to Problem 10.99(a) is easily extended to show that

$$\Psi(x, x^{1/u}) = (1 - \log u)x + O(x/\log x), \tag{10.28}$$

for $x > 1$ and $1 \leq u \leq 2$. This is good as far as it goes, but we would also like to also understand $\Psi(x, y)$ when y is smaller than $\sqrt{x}$, perhaps significantly smaller.

To start the discussion, suppose that $u \in [2, 3]$, and notice that

$$\Psi(x, x^{1/u}) = \Psi(x, x^{1/2}) - \sum_{x^{1/u} < p \leq x^{1/2}} \Psi(x/p, p).$$

Indeed, on the right-hand side, the first term counts all $x^{1/2}$-smooth numbers up to x, while in the sum we successively remove, for each p between $x^{1/u}$ and $x^{1/2}$, the count of numbers with largest prime factor p. The numbers that are still counted are precisely the $x^{1/u}$-smooth integers up to x. What have we gained with this identity? For every term in the right-hand sum, $p \geq (x/p)^{1/2}$, so we can estimate $\Psi(x/p, p)$ using (10.28). With some effort, this idea leads to the result that for $u \in [2, 3]$,

$$\Psi(x, x^{1/u}) = \left((1 - \log 2) - \int_2^u \frac{1 - \log(t - 1)}{t} \, dt \right) x + O(x/\log x).$$

One can continue this process. In general, if one understands the behavior of $\Psi(x, x^{1/u})$ when $u \in [1, k]$ (for a fixed integer $k \geq 1$), then one can understand $\Psi(x, x^{1/u})$ for $u \in [k, k+1]$ via the identity

$$\Psi(x, x^{1/u}) = \Psi(x, x^{1/k}) - \sum_{x^{1/u} < p \leq x^{1/k}} \Psi(x/p, p).$$

Filling in the details, one is led to the following theorem, a form of which can be found in Ramanujan's 'Lost Notebook' (see [1, Chapter 8]): Let $\rho(u) = 1 - \log u$ for $1 \leq u \leq 2$, and successively define, on $[k, k+1] = [2, 3], [3, 4], [4, 5]$, etc.,

$$\rho(u) = \rho(k) - \int_k^u \frac{\rho(t-1)}{t} \, dt.$$

For any positive integer k, any x > 1, and any u with $1 \le u \le k$,

$$\Psi(x, x^{1/u}) = x\rho(u) + O(x/\log x), \tag{10.29}$$

where the constant implied by the O-symbol depends only on k.

In view of the relation (10.29), it is natural to ask how much we know about the values of $\rho(u)$, as u varies. There is no closed form expression for $\rho(u)$ valid for all u, but one can obtain fairly precise asymptotic results. For instance, it can be proved that $\log \rho(u) = -u \log u - u \log \log u + u + O(u \log \log u/\log u)$, as $u \to \infty$, so that $\rho(u) \approx (\frac{e}{u \log u})^u$ for large values of u.

There is much more to say about $\Psi(x, y)$. For instance, in many applications it is important to understand $\Psi(x, y)$ when $x, y \to \infty$ and y is smaller than any fixed power of x (for example, when $y = \log x$). This range of x and y is left untouched by our arguments above. In fact, the study of $\Psi(x, y)$ ("psixyology") could occupy a monograph by itself. We stop here and refer the interested reader to the surveys of Granville [2] and Moree [6].

10.100

(a) Writing $\omega(n) = \sum_{p|n} 1$, we see that

$$\sum_{n\le x} \omega(n) = \sum_{n\le x} \sum_{p|n} 1 = \sum_{p\le x} \sum_{\substack{n\le x \\ p|n}} 1 = \sum_{p\le x} \left\lfloor \frac{x}{p} \right\rfloor = x \sum_{p\le x} \frac{1}{p} + O(\pi(x))$$

$$= x(\log \log x + O(1)) + O(x) = x \log \log x + O(x).$$

Here $\sum_{p\le x} \frac{1}{p}$ was estimated using Problem 10.98.

(b) Writing $\omega(n)^2 = \sum_{p,q|n} 1$, as in the hint,

$$\sum_{n\le x} \omega(n)^2 = \sum_{p,q\le x} \sum_{\substack{n\le x \\ p,q|n}} 1 = \sum_{p\le x} \left\lfloor \frac{x}{p} \right\rfloor + \sum_{\substack{p,q\le x \\ p\ne q}} \left\lfloor \frac{x}{pq} \right\rfloor.$$

As we saw in (a), $\sum_{p\le x} \lfloor \frac{x}{p} \rfloor = O(x \log \log x)$. Also,

$$\sum_{\substack{p,q\le x \\ p\ne q}} \left\lfloor \frac{x}{pq} \right\rfloor = \sum_{\substack{p,q \\ p\ne q,\ pq\le x}} \left\lfloor \frac{x}{pq} \right\rfloor = x \sum_{\substack{p,q \\ p\ne q,\ pq\le x}} \frac{1}{pq} + O\left(\sum_{\substack{p,q \\ pq\le x}} 1 \right).$$

The error term here is $O(x)$, since each integer can be written in at most two ways as pq with p, q prime. Moreover, since $\sum_p \frac{1}{p^2} < \infty$,

$$\sum_{\substack{p,q \\ p\ne q,\ pq\le x}} \frac{1}{pq} = \sum_{\substack{p,q \\ pq\le x}} \frac{1}{pq} + O(1).$$

Collecting estimates, $\sum_{n \le x} \omega(n)^2 = x \sum_{\substack{p,q \\ pq \le x}} \frac{1}{pq} + O(x \log \log x)$, as desired.

(c) The key is to recognize that

$$\left(\sum_{p \le \sqrt{x}} \frac{1}{p} \right)^2 \le \sum_{\substack{p,q \\ pq \le x}} \frac{1}{pq} \le \left(\sum_{p \le x} \frac{1}{p} \right)^2.$$

Both $\sum_{p \le \sqrt{x}} \frac{1}{p}$ and $\sum_{p \le x} \frac{1}{p}$ are $\log \log x + O(1)$, and so when squared are $(\log \log x)^2 + O(\log \log x)$. It follows that

$$\sum_{\substack{p,q \\ pq \le x}} \frac{1}{pq} = (\log \log x)^2 + O(\log \log x).$$

Plugging this into (b), $\sum_{n \le x} \omega(n)^2 = x(\log \log x)^2 + O(x \log \log x)$.

10.101 It is easy to convince oneself that $\Omega(n) = \sum_{p^k | n, \, k \ge 1} 1$, so that $\Omega(n) - \omega(n) = \sum_{p^k | n, \, k \ge 2} 1$. Hence,

$$0 \le \sum_{n \le x} (\Omega(n) - \omega(n)) = \sum_{\substack{p^k \le x \\ k \ge 2}} \sum_{\substack{n \le x \\ p^k | n}} 1 \le \sum_{\substack{p^k \le x \\ k \ge 2}} \frac{x}{p^k}$$

$$\le x \sum_p \sum_{k \ge 2} \frac{1}{p^k} = x \sum_p \frac{1}{p^2 - p} < x \sum_{m \ge 2} \frac{1}{m^2 - m} = x.$$

10.102 Recall that

$$\frac{n}{\phi(n)} = \prod_{p | n} \left(1 - \frac{1}{p} \right)^{-1}. \tag{10.30}$$

We consider separately the contribution to the product from the terms with $p \le \log n$ versus $p > \log n$.

It is easily proved by induction that $\prod_i (1 - a_i) \ge 1 - \sum_i a_i$, for any finite sequence of a_i in $[0, 1]$. Thus,

$$\prod_{\substack{p | n \\ p > \log n}} \left(1 - \frac{1}{p} \right) \ge 1 - \sum_{\substack{p | n \\ p > \log n}} \frac{1}{p} \ge 1 - \frac{1}{\log n} \cdot \omega_{> \log n}(n),$$

where $\omega_{> \log n}(n)$ denotes the number of prime factors of n exceeding $\log n$. Noting that $n \ge \prod_{p | n, \, p > \log n} p \ge (\log n)^{\omega_{> \log n}(n)}$, we deduce that $\omega_{> \log n}(n) \le \frac{\log n}{\log \log n}$.

Putting this back above, we see that for large n,

$$\prod_{\substack{p|n \\ p>\log n}} \left(1-\frac{1}{p}\right) \geq 1 - \frac{1}{\log\log n} > \frac{1}{2},$$

and

$$\prod_{\substack{p|n \\ p>\log n}} \left(1-\frac{1}{p}\right)^{-1} \leq 2. \tag{10.31}$$

Turning now to the primes $p \leq \log n$, clearly

$$\prod_{\substack{p|n \\ p\leq\log n}} \left(1-\frac{1}{p}\right)^{-1} \leq \prod_{p\leq\log n} \left(1-\frac{1}{p}\right)^{-1}$$

$$= \prod_{p\leq\log n} \left(1+\frac{1}{p-1}\right) \leq \exp\left(\sum_{p\leq\log n}\frac{1}{p-1}\right). \tag{10.32}$$

Moreover,

$$\sum_{p\leq\log n}\frac{1}{p-1} \leq \sum_{p\leq\log n}\frac{1}{p} + \sum_{p}\frac{1}{p(p-1)} \leq \log\log\log n + O(1),$$

so that

$$\exp\left(\sum_{p\leq\log n}\frac{1}{p-1}\right) = O(\log\log n). \tag{10.33}$$

The desired upper bound follows from combining (10.30)–(10.33).

10.103

(a) Since $o(2 \bmod \ell)$ is a proper divisor of n, there is at least one prime p dividing n to a higher power than that to which it divides $o(2 \bmod \ell)$. Any such prime p has the desired property.

(b) Let p be a prime satisfying (a), meaning that $p \mid n$ and $o(2 \bmod \ell)$ divides n/p. Then ℓ divides $2^{n/p} - 1$, while also $\ell \mid \frac{2^n-1}{2^{n/p}-1}$ from the definition of B_n. So working modulo ℓ,

$$0 \equiv \frac{2^n - 1}{2^{n/p} - 1}$$

$$= (2^{n/p})^{p-1} + (2^{n/p})^{p-2} + \cdots + 2^{n/p} + 1$$

$$\equiv \underbrace{1 + \cdots + 1}_{p \text{ times}} \equiv p.$$

This forces $\ell = p$. So the prime p in (a) is unique, and $p = \ell$. Note that as a byproduct we get that $\ell \mid n$.

(c) Let q be a prime dividing n with $q \neq \ell$. Then q does not satisfy the conclusion of (a). Hence, $o(2 \bmod \ell)$ does not divide n/q. But $o(2 \bmod \ell)$ does divide n; the only way this is possible is if $q \mid o(2 \bmod \ell)$.

(d) From (c), if q is a prime dividing n with $q \neq \ell$, then $q \mid o(2 \bmod \ell)$. But $o(2 \bmod \ell) \mid \ell - 1$, so that $q < \ell$. Since ℓ is a prime dividing n and all other prime divisors of n are less than ℓ, we conclude that ℓ is the largest prime factor of n.

(e) It is enough to show that $\ell^2 \nmid \frac{2^n-1}{2^{n/\ell}-1}$. Write $2^{n/\ell} - 1 = \ell^r q$, where $r \geq 1$ and $\ell \nmid q$. Note, as will be useful momentarily, that ℓ is odd and so $\ell \geq 3$. Then

$$2^n = (1 + \ell^r q)^\ell = 1 + \ell^{r+1} q + \sum_{j \geq 2} \binom{\ell}{j} \ell^{jr} q^j.$$

All of the terms in the sum on j are divisible by ℓ^{2r+1}: When $2 \leq j < \ell$, one sees this by noting that ℓ^{2r} divides ℓ^{jr}, while $\ell \mid \binom{\ell}{j}$. When $j = \ell$, the corresponding term is divisible by ℓ^{3r} (since $\ell \geq 3$), and $3r \geq 2r + 1$. Thus,

$$2^n - 1 = \ell^{r+1} \Bigg(\underbrace{q}_{\not\equiv 0 \ (\bmod \ \ell)} + \overbrace{\sum_{j \geq 2} \binom{\ell}{j} \ell^{jr-r-1} q^j}^{\equiv 0 \ (\bmod \ \ell)} \Bigg),$$

and the parenthesized factor is not a multiple of ℓ. Hence, ℓ^{r+1} is the largest power of ℓ dividing $2^n - 1$, and $\ell^2 \nmid \frac{2^n-1}{2^{n/\ell}-1}$.

10.104 We show that if n is repellent and $B_n > 1$, then B_n is the largest prime factor of n. Let ℓ be any prime dividing B_n. By assumption, $o(2 \bmod \ell) < n$ and so, from the last problem, ℓ is the largest prime factor of n. Since this holds for every prime dividing B_n, it must be that B_n is a power of ℓ. But we know that $\ell^2 \nmid B_n$, and so $B_n = \ell$.

In the table of values of B_n for $2 \leq n \leq 18$, we see that the only time B_n is 1 or the largest prime factor of n is $n = 6$. And 6 is indeed repellent: If ℓ is a prime with $o(2 \bmod \ell) = 6$, then $\ell \mid B_6 = 3$, so we need $\ell = 3$—but $o(2 \bmod 3) = 2$.

10.105 Recall that in Problem 7.74, you showed that

$$\sum_{i=1}^{k-1} \zeta(2i)\zeta(2k-2i)$$

$$= (k-1)\zeta(2k) + \sum_{\substack{m,n\geq 1 \\ m\neq n}} \left(\frac{1}{m^{2k-2}}\frac{1}{n^2-m^2} + \frac{1}{n^{2k-2}}\frac{1}{m^2-n^2} \right).$$

The double sum is symmetric in m and n, which suggests that

$$\sum_{\substack{m,n\geq 1 \\ m\neq n}} \left(\frac{1}{m^{2k-2}}\frac{1}{n^2-m^2} + \frac{1}{n^{2k-2}}\frac{1}{m^2-n^2} \right) = 2\sum_{m=1}^{\infty} \sum_{\substack{n\in\mathbb{Z}^+ \\ n\neq m}} \frac{1}{m^{2k-2}}\frac{1}{n^2-m^2}.$$

(10.34)

We will give a rigorous justification of (10.34) at the end of this solution. Assuming this for now, we observe that by Problem 9.95,

$$2\sum_{m=1}^{\infty} \sum_{\substack{n\in\mathbb{Z}^+ \\ n\neq m}} \frac{1}{m^{2k-2}}\frac{1}{n^2-m^2} = 2\sum_{m=1}^{\infty} \frac{1}{m^{2k-2}} \cdot \frac{3}{4m^2} = \frac{3}{2}\zeta(2k).$$

Substituting this above,

$$\sum_{i=1}^{k-1} \zeta(2i)\zeta(2k-2i) = \left(k + \frac{1}{2}\right)\zeta(2k),$$

which is the desired identity.

Plugging $k = 2$ into our identity shows that $\zeta(2)^2 = \frac{5}{2}\zeta(4)$. Recalling that $\zeta(2) = \frac{\pi^2}{6}$, we infer that $\zeta(4) = \frac{\pi^4}{90}$.

The proof that $\zeta(2k) \in \pi^{2k}\mathbb{Q}$ is a straightforward argument by complete induction. The claim holds when $k = 1$. If it holds for all positive integers $< k$, then $\zeta(2k-2i)\zeta(2i) \in (\pi^{2k-2i}\mathbb{Q})(\pi^{2i}\mathbb{Q}) \subseteq \pi^{2k}\mathbb{Q}$ for all $i = 1, 2, \ldots, k-1$. Hence, $(k + \frac{1}{2})\zeta(2k) \in \pi^{2k}\mathbb{Q}$, and so $\zeta(2k) \in \pi^{2k}\mathbb{Q}$.

What about (10.34)? A naive argument for (10.34) would proceed in two steps. First, splitting the sum into obvious "halves" and exploiting the symmetry,

$$\sum_{\substack{m,n\geq 1 \\ m\neq n}} \left(\frac{1}{m^{2k-2}}\frac{1}{n^2-m^2} + \frac{1}{n^{2k-2}}\frac{1}{m^2-n^2} \right) = 2\sum_{\substack{m,n\geq 1 \\ m\neq n}} \frac{1}{m^{2k-2}}\frac{1}{n^2-m^2}.$$

(10.35)

Now rewriting the right-hand double sum as an iterated sum, with the sum on m outside, gives (10.34).

The problem with this "proof" is that the terms $\frac{1}{m^{2k-2}}\frac{1}{n^2-m^2}$ are not all positive. This renders both of the above steps somewhat suspect. All such concerns vanish if we show that the right-hand sum in (10.35) converges absolutely; that is,

$$\sum_{\substack{m,n\geq 1 \\ m\neq n}} \frac{1}{m^{2k-2}} \frac{1}{|n^2-m^2|} < \infty. \tag{10.36}$$

Thankfully, (10.36) is not so difficult. Since $\sum_{m\geq 1} \frac{1}{m^{2k-2}} < \infty$, it suffices to prove that for every positive integer m, the sum $\sum_{n\in\mathbb{Z}^+,\, n\neq m} \frac{1}{|n^2-m^2|} = O(1)$, with an implied constant independent of m. This sum on n has $m-1$ terms corresponding to $n < m$, and for each of these, $\frac{1}{|n^2-m^2|} \leq \frac{1}{m^2-(m-1)^2} = \frac{1}{2m-1}$, so these terms contribute at most $\frac{m-1}{2m-1} < 1$. Now suppose $n > m$, say $n = m+r$. Then $n^2-m^2 = 2mr+r^2 \geq r^2$, and so the terms with $n > m$ contribute at most $\sum_r \frac{1}{r^2} < 2$. Hence, $\sum_{n\in\mathbb{Z}^+,\, n\neq m} \frac{1}{|n^2-m^2|} < 3$, for every m.

Remarks.

(a) It was known already to Euler that the values $\zeta(2k)$ appear in the coefficients of the power series expansion of cotangent. Specifically, for x near 0,

$$\frac{1}{2} - \frac{1}{2}\pi x \cot(\pi x) = \sum_{k=1}^{\infty} \zeta(2k)x^{2k}. \tag{10.37}$$

Euler's proof of (10.37) starts by establishing the beautiful (and independently interesting) identity

$$\sin(\pi x) = \pi x \prod_{n\geq 1}\left(1 - \frac{x^2}{n^2}\right), \tag{10.38}$$

for all real numbers x. This approach to (10.37) can be found in many textbooks. Usually the product formula (10.38) is proved using ideas from complex analysis, but this can be avoided, as shown (for example) by Koblitz in [5].

Alternatively, the identity (10.37) can be deduced from Problem 10.105 coupled with our previous determination of $\zeta(2)$. For x sufficiently close to 0,

$$\frac{1}{2} - \frac{1}{2}\pi x \cot(\pi x) = \frac{1}{2} - \frac{1}{2}\pi x \left(\frac{1 - \frac{\pi^2 x^2}{2!} + \frac{\pi^4 x^4}{4!} - \frac{\pi^6 x^6}{6!} + \cdots}{\pi x - \frac{\pi^3 x^3}{3!} + \frac{\pi^5 x^5}{5!} - \frac{\pi^7 x^7}{7!} + \cdots}\right)$$

$$= \sum_{k\geq 1} c_{2k} x^{2k}, \quad \text{for certain real numbers } c_{2k}. \tag{10.39}$$

By a direct calculation, $c_2 = \frac{\pi^2}{6}$, which we recognize as $\zeta(2)$. Now differentiate both sides of (10.39) and multiply by x. Since $\cot' = -1 - \cot^2$, one can use (10.39) to determine a power series expansion of the resulting left-hand side. The resulting right-hand side is, of course, simply $\sum_{k\geq 1} 2k \cdot c_{2k} x^{2k}$. Comparing coefficients of x^{2k}, one finds after a little computation that $(k + \frac{1}{2})c_{2k} = \sum_{i=1}^{k-1} c_{2i} c_{2k-2i}$ for $k \geq 2$. Since c_{2k} and $\zeta(2k)$ satisfy the same recurrence, and agree for $k = 1$, we conclude that $c_{2k} = \zeta(2k)$ for all k.

(b) A celebrated 1882 theorem of Lindemann asserts that π is "transcendental", that is, not a root of a nonzero polynomial with rational coefficients. (We recommend Hadlock's book [3] for a readable account of Lindemann's theorem.) This result together with Problem 10.105 implies that $\zeta(2k)$ is also transcendental, for every positive integer k.

Much less is known about the values $\zeta(2k + 1)$. It was only in 1978 that Apéry succeeded in proving the irrationality of $\zeta(3)$ (irrationality being a much weaker property than transcendence), and it is still an open problem to decide the irrationality of $\zeta(5)$. There are some glimmers of hope, though: Ball and Rivoal have shown that $\zeta(2k + 1)$ is irrational for infinitely many k, and Zudilin has shown that at least one of $\zeta(5), \zeta(7), \zeta(9), \zeta(11)$ is irrational.

(c) Problems 7.74, 9.95, and 10.105 were based on Murty's article [7].

10.106

(a) Suppose n is odd and perfect, and factor n into primes, say $n = \prod_{p|n} p^{e_p}$. Then

$$2n = \sigma(n) = \prod_{p|n} \sigma(p^{e_p}).$$

Since $2n$ is twice an odd number, $\sigma(p^{e_p})$ is odd for every prime power p^{e_p} appearing in the prime factorization of n, with precisely one exception. But for each odd prime p and positive integer e,

$$\sigma(p^e) = 1 + p + \cdots + p^e \equiv e + 1 \pmod{2}.$$

It follows that e_p is even for every prime p dividing n, with exactly one exception. The claim in (a) follows with $m = \prod_{p|n,\ 2|e_p} p^{\frac{1}{2}e_p}$.

(b) Suppose that n is an odd perfect number with $n \le x$, and write $n = p^k m^2$ as in (a). Then $m \le \sqrt{x}$, and so it suffices to prove that m determines n. To this end, observe that

$$2p^k m^2 = 2n = \sigma(n) = \sigma(p^k)\sigma(m^2),$$

so that

$$\frac{2m^2}{\sigma(m^2)} = \frac{\sigma(p^k)}{p^k}.$$

The right-hand fraction is reduced, as its numerator $1 + p + \cdots + p^k$ is coprime to p. Since each rational number has a unique lowest-terms representation, p^k, and hence also $n = p^k m^2$, is determined by m: Namely, p^k is the denominator when $2m^2/\sigma(m^2)$ is reduced.

Remark. In 1957, 2 years after Hornfeck proved the $x^{1/2}$ upper bound of Problem 10.106(b), Hornfeck and Wirsing [4] proved that the count of odd perfect numbers up to x is at most x^ϵ, for any fixed $\epsilon > 0$ and all large x. It is widely believed that there are no odd perfect numbers, so that the "correct" upper bound in this problem is not $x^{1/2}$ or x^ϵ, but 0 ! But that goal appears rather distant. It would seem to require new ideas to prove, for example, that the count of odd perfect numbers up to x is $O((\log x)^{10^{10}})$, as $x \to \infty$.

References

1. George E. Andrews and Bruce C. Berndt, *Ramanujan's lost notebook. Part IV.* Springer, New York, 2013.
2. Andrew Granville, *Smooth numbers: computational number theory and beyond.* Algorithmic number theory: lattices, number fields, curves and cryptography, 267–323, Math. Sci. Res. Inst. Publ., vol. 44, Cambridge Univ. Press, Cambridge, 2008.
3. Charles Robert Hadlock, *Field theory and its classical problems.* Carus Mathematical Monographs, vol. 19. Mathematical Association of America, Washington, D.C., 1978.
4. Bernhard Hornfeck and Eduard Wirsing, *Über die Häufigkeit vollkommener Zahlen.* Math. Ann. 133 (1957), 431–438.
5. Neal Koblitz, *p-adic numbers, p-adic analysis, and zeta-functions.* Second edition. Graduate Texts in Mathematics, vol. 58. Springer-Verlag, New York, 1984.
6. Pieter Moree, *Integers without large prime factors: from Ramanujan to de Bruijn.* Integers 14A (2014), Paper No. A5, 13 pp.
7. M. Ram Murty, *A simple proof that* $\zeta(2) = \frac{\pi^2}{6}$. Math. Student 88 (2019), 113–115.

Solutions to Step #11

11.107 Note that $p^{\frac{1}{3}} > (p^{\frac{1}{2}+\epsilon})^{\frac{1}{2}}$. Now the same reasoning as in the solution to Problem 10.99 shows that the limiting proportion of $n \le p^{\frac{1}{2}+\epsilon}$ possessing a prime factor $> p^{\frac{1}{3}}$ is

$$\log\log p^{\frac{1}{2}+\epsilon} - \log\log p^{\frac{1}{3}} = \log\frac{\log p^{\frac{1}{2}+\epsilon}}{\log p^{\frac{1}{3}}} = \log\frac{1/2+\epsilon}{1/3}$$

$$< \log\frac{1/2+1/50}{1/3} = 0.444\cdots < 0.45.$$

11.108 Let $\epsilon \in (0, 1/50)$. By Exercise 6.56, once p is large enough at least 49% of the integers in $[1, p^{\frac{1}{2}+\epsilon}]$ are nonsquares modulo p. Fewer than 45% of the integers from the same interval have a prime factor $> p^{1/3}$. So there is integer $n \in [1, p^{\frac{1}{2}+\epsilon}]$ that is both a nonsquare modulo p and factors completely into primes not exceeding $p^{\frac{1}{3}}$. Not every prime factor of n can be a square, otherwise n itself would be a square. So there is a prime $< p^{\frac{1}{3}}$ reducing to a nonsquare modulo p.

11.109 Fix $\epsilon \in (0, 1/2)$. We look at the integers in $[1, p^{\frac{1}{2}+\epsilon}]$ which have a prime factor $> p^{\frac{1}{2\sqrt{e}}+\epsilon}$. The limiting proportion of these integers (as $p \to \infty$) is

$$\eta := \log\frac{\frac{1}{2}+\epsilon}{\frac{1}{2\sqrt{e}}+\epsilon} < \frac{1}{2}.$$

The solution of Problem 6.56 shows that for all large enough p, the proportion of nonsquares mod p among the integers in $[1, p^{\frac{1}{2}+\epsilon}]$ is larger than η. Thus, there is a

© Springer Nature Switzerland AG 2021
P. Pollack, A. Singha Roy, *Steps into Analytic Number Theory*, Problem Books in
Mathematics, https://doi.org/10.1007/978-3-030-65077-3_26

nonsquare in $[1, p^{\frac{1}{2}+\epsilon}]$ factoring entirely into primes $< p^{\frac{1}{2\sqrt{e}}+\epsilon}$. As a consequence, there is some prime $< p^{\frac{1}{2\sqrt{e}}+\epsilon}$ reducing to a nonsquare mod p.

11.110 Clearly,

$$\sum_{n\le x}(\omega(n) - \log\log x)^2 = \sum_{n\le x}\omega(n)^2 - 2\log\log x\sum_{n\le x}\omega(n) + (\log\log x)^2\sum_{n\le x}1.$$

By Problem 10.100,

$$\sum_{n\le x}\omega(n)^2 = x(\log\log x)^2 + O(x\log\log x),$$

$$2\log\log x\sum_{n\le x}\omega(n) = 2x(\log\log x)^2 + O(x\log\log x),$$

$$(\log\log x)^2\sum_{n\le x}1 = x(\log\log x)^2 + O((\log\log x)^2).$$

Substituting these expressions above and simplifying,

$$\sum_{n\le x}(\omega(n) - \log\log x)^2 = O(x\log\log x).$$

11.111 Fix $\epsilon > 0$. Let N be the number of $n \le x$ with $|\omega(n) - \epsilon\log\log x| > \epsilon\log\log x$, so that $N/\lfloor x\rfloor$ represents the relative frequency of such $n \le x$.

For every $n \in \mathbb{Z}^+$, we have $(\omega(n) - \log\log x)^2 \ge 0$. Moreover, if $|\omega(n) - \log\log x| > \epsilon\log\log x$, then $(\omega(n) - \log\log x)^2 > \epsilon^2(\log\log x)^2$. Therefore,

$$\epsilon^2(\log\log x)^2 N \le \sum_{n\le x}(\omega(n) - \log\log x)^2 = O(x\log\log x).$$

Hence, $N = O(\epsilon^{-2}x/\log\log x)$, and the ratio $N/\lfloor x\rfloor \to 0$ as $x \to \infty$.

Remark. Those who have seen some probability theory will recognize this proof as an application of "Chebyshev's inequality".

11.112 Fix $\epsilon > 0$. Let N, N', and N'' denote (respectively) the counts of $n \le x$ with

$$|\Omega(n) - \log\log x| > \epsilon\log\log x,$$

$$|\omega(n) - \log\log x| > \frac{1}{2}\epsilon\log\log x, \text{ and}$$

$$|\Omega(n) - \omega(n)| > \frac{1}{2}\epsilon\log\log x.$$

Then

$$N \leq N' + N''.$$

From Problem 11.111, we have that $N'/\lfloor x \rfloor \to 0$ (as $x \to \infty$). Moreover, from Problem 10.101,

$$N'' \cdot \frac{1}{2} \epsilon \log \log x \leq \sum_{n \leq x} |\Omega(n) - \omega(n)|$$

$$= \sum_{n \leq x} (\Omega(n) - \omega(n)) = O(x).$$

Thus, $N'' = O(\epsilon^{-1} x / \log \log x)$ and $N''/\lfloor x \rfloor \to 0$. We conclude that $N/\lfloor x \rfloor \to 0$ as well.

11.113 Fix $\epsilon > 0$. By Problems 11.111 and 11.112, we have that

$$(1 - \epsilon) \log \log x \leq \omega(n), \quad \Omega(n) \leq (1 + \epsilon) \log \log x$$

for all $n \leq x$, apart from an exceptional set of n with limiting frequency 0. It now suffices to observe that for all $n \leq x$ not in this exceptional set,

$$\tau(n) \geq 2^{\omega(n)} \geq 2^{(1-\epsilon) \log \log x} = (\log x)^{(\log 2)(1-\epsilon)} > (\log x)^{\log 2 - \epsilon},$$

and

$$\tau(n) \leq 2^{\Omega(n)} \leq 2^{(1+\epsilon) \log \log x} = (\log x)^{(\log 2)(1+\epsilon)} < (\log x)^{\log 2 + \epsilon}.$$

Here we used the bounds for $\tau(n)$ proved in Problem 5.49.

Since $\log 2 < 1$, a typical $n \leq x$ has many fewer divisors than the average of $\approx \log x$.

Remark. How can it be that the average order of $\tau(n)$ is so much larger than its typical order? The explanation is that the average is dominated by a relatively sparse set of n on which $\tau(n)$ is much larger than usual. While a typical $n \leq x$ has $\omega(n) \approx \log \log x$, it can be shown that the average of $\tau(n)$ on $n \leq x$ is essentially determined by those n with $\omega(n) \approx 2 \log \log x$: *For each $\epsilon > 0$, there is a $\delta > 0$ such that, for all $x > x_0(\epsilon)$,*

$$\sum_{\substack{n \leq x \\ |\omega(n) - 2 \log \log x| > \epsilon \log \log x}} \tau(n) < x (\log x)^{1-\delta}.$$

See [1] for a general theorem containing this and many similar results.

11.114

(a) Write $A = \prod_{p \le \log n / (\log \log n)^3} p^{e_p}$. Since A divides n, it is certainly the case that each $p^{e_p} \le n$, and so each $e_p \le \frac{\log n}{\log 2}$. Thus, for large enough n,

$$\tau(A) = \prod_{p \le \log n / (\log \log n)^3} (e_p + 1)$$

$$\le \left(\frac{\log n}{\log 2} + 1 \right)^{\log n / (\log \log n)^3}$$

$$= \exp \left(\frac{\log n}{(\log \log n)^3} \cdot \log \left(\frac{\log n}{\log 2} + 1 \right) \right)$$

$$< \exp \left(\frac{\log n}{(\log \log n)^3} \cdot 2 \log \log n \right)$$

$$= \exp \left(2 \cdot \frac{\log n}{(\log \log n)^2} \right).$$

(b) Since every prime dividing B exceeds $\frac{\log n}{(\log \log n)^3}$,

$$n \ge \left(\frac{\log n}{(\log \log n)^3} \right)^{\Omega(B)},$$

so that for large n,

$$\Omega(B) \le \frac{\log n}{\log \frac{\log n}{(\log \log n)^3}}$$

$$= \frac{\log n}{\log \log n - 3 \log \log \log n}$$

$$= \frac{\log n}{\log \log n} \left(\frac{1}{1 - \frac{3 \log \log \log n}{\log \log n}} \right)$$

$$< \frac{\log n}{\log \log n} \left(1 + 6 \frac{\log \log \log n}{\log \log n} \right)$$

$$= \frac{\log n}{\log \log n} + 6 \frac{\log n \cdot \log \log \log n}{(\log \log n)^2}.$$

The bound on $\tau(B)$ now follows from Problem 5.49.

(c) From (a) and (b), we have for large n that

$$\tau(n) = \tau(AB) = \tau(A)\tau(B)$$

$$\leq 2^{\frac{\log n}{\log \log n} + O\left(\frac{\log n \cdot \log \log \log n}{(\log \log n)^2}\right)},$$

and the exponent here is eventually smaller than $(1 + \epsilon)\frac{\log n}{\log \log n}$.

(d) We start by noting that $\tau(\mathrm{LCM}(x)) \geq 2^{\omega(\mathrm{LCM}(x))} = 2^{\pi(x)}$. So it suffices to show that for each fixed $\epsilon > 0$, and all large x,

$$\pi(x) > (1 - \epsilon)\frac{\log \mathrm{LCM}(x)}{\log \log \mathrm{LCM}(x)}.$$

We will prove this in the equivalent form

$$(1 - \epsilon)\log \mathrm{LCM}(x) < \pi(x) \cdot \log \log \mathrm{LCM}(x).$$

Taking $2n + 1$ as the largest odd integer not exceeding x, we see from Problem 7.69 that for large x,

$$\mathrm{LCM}(x) \geq \mathrm{LCM}(2n + 1) \geq 4^n > 4^{x/3}.$$

Thus,

$$\log \log \mathrm{LCM}(x) \geq \log x + \log \frac{\log 4}{3} > (1 - \epsilon)\log x$$

and

$$\pi(x) \log \log \mathrm{LCM}(x) > (1 - \epsilon)\pi(x) \log x \geq (1 - \epsilon)\log \mathrm{LCM}(x),$$

using in the last step the result of Exercise 8.79.

Remark. A sharper result on the maximal size of $\tau(n)$ was obtained by Ramanujan in [3]. Fix any nonnegative integer k. Then, for all $n \geq 3$,

$$\frac{\log \tau(n)}{\log 2} \leq \frac{\log n}{\log \log n} \sum_{j=0}^{k} \frac{j!}{(\log \log n)^j} + O_k\left(\frac{\log n}{(\log \log n)^{k+2}}\right); \tag{11.40}$$

moreover, equality holds here along a sequence of n tending to infinity. Note that taking $k = 0$ in (11.40) already gives a more precise upper bound for $\tau(n)$ than the one proved in part (c).

11.115 It suffices to show that $\Phi_n(2) \mid \frac{2^n - 1}{2^{n/p} - 1}$ for each prime divisor p of n. We know that

$$\prod_{d \mid n} \Phi_d(2) = 2^n - 1, \quad \text{while} \quad \prod_{d \mid n/p} \Phi_d(2) = 2^{n/p} - 1.$$

Hence,

$$\frac{2^n - 1}{2^{n/p} - 1} = \prod_{\substack{d|n \\ d\nmid n/p}} \Phi_d(2).$$

All of the terms in the right-hand product are integers, and one of them is $\Phi_n(2)$. The result follows.

Remark. With more effort, one can show that $B_n = \Phi_n(2)$. Compare with the remark on p. 6 of [5].

11.116 We write

$$\Phi_n(2) = \prod_{d|n}(2^d - 1)^{\mu(n/d)} = \prod_{d|n}(2^d)^{\mu(n/d)} \prod_{d|n}\left(1 - \frac{1}{2^d}\right)^{\mu(n/d)}.$$

Notice that

$$\prod_{d|n}(2^d)^{\mu(n/d)} = 2^{\sum_{d|n} d\mu(n/d)} = 2^{\sum_{d|n}(n/d)\mu(d)} = 2^{\phi(n)},$$

and so we have the equality claimed in the problem. To show the inequality, it suffices to prove that

$$\prod_{d|n}\left(1 - \frac{1}{2^d}\right)^{\mu(n/d)} > \frac{1}{3}.$$

We take cases. First, if $\mu(n) = 0$ or -1, then

$$\prod_{d|n}\left(1 - \frac{1}{2^d}\right)^{\mu(\frac{n}{d})} \geq \prod_{\substack{d|n \\ d>1}}\left(1 - \frac{1}{2^d}\right)^{\mu(\frac{n}{d})} \geq \prod_{d>1}\left(1 - \frac{1}{2^d}\right) \geq 1 - \sum_{d>1}\frac{1}{2^d} = \frac{1}{2},$$

so we may assume that $\mu(n) = 1$. If $n = 1$, the result is easily verified, so we also assume that $n > 1$. Let p be the smallest prime dividing n. If $d > 1$ and d divides n, then $d \geq p$. Moreover, when $d = p$, we have $\mu(n/d) = -1$, so that the $d = p$ term makes a contribution of size > 1 to our product. We conclude that

$$\prod_{d|n}\left(1 - \frac{1}{2^d}\right)^{\mu(\frac{n}{d})} > \left(1 - \frac{1}{2}\right)\prod_{d>p}\left(1 - \frac{1}{2^d}\right)$$

$$\geq \frac{1}{2}\left(1 - \sum_{d>p}\frac{1}{2^d}\right) = \frac{1}{2}\left(1 - \frac{1}{2^p}\right) \geq \frac{1}{2}\left(1 - \frac{1}{4}\right) > \frac{1}{3}.$$

Remark. In fact, $\Phi_n(2) \geq \frac{1}{2}2^{\phi(n)}$ for all n (see [2] for much more general results).

11.117 If $n > 1$ and there is no prime ℓ with $o(2 \bmod \ell) = n$, then $B_n = 1$ or p, where p is the largest prime factor of n (Problem 10.104). Thus, it will suffice to prove that $B_n > p$ for every integer $n > 6$.

Since $p - 1 \mid \phi(n)$, we get from Problems 11.115 and 11.116 that

$$B_n \geq \Phi_n(2) > \frac{1}{3} 2^{\phi(n)} \geq \frac{1}{3} 2^{p-1},$$

which is larger than p once $p \geq 5$. So we can assume $p = 2$ or $p = 3$. If $p = 2$, then $n = 2^k$ with $k \geq 3$ (we are assuming $n \geq 6$), and

$$B_n \geq \Phi_{2^k}(2) = \frac{2^{2^k} - 1}{2^{2^{k-1}} - 1} = 2^{2^{k-1}} + 1 > 2.$$

So suppose now that $p = 3$. If $n = 3^k$, then $k \geq 2$, and

$$B_n \geq \Phi_{3^k}(2) = \frac{2^{3^k} - 1}{2^{3^{k-1}} - 1} = 2^{2 \cdot 3^{k-1}} + 2^{3^{k-1}} + 1 > 3.$$

It remains to consider those cases when $p = 3$ and $6 \mid n$. Write $n = 6m$, where no prime greater than 3 divides m. Since $n > 6$, we have $m \geq 2$. Then

$$B_n \geq \Phi_n(2) > \frac{1}{3} 2^{\phi(n)} = \frac{1}{3} 2^{n/3} = \frac{1}{3} 2^{2m} \geq \frac{1}{3} 2^4 > 3.$$

Therefore, $B_n > p$ for all $n > 6$.

Remark. One can push these same methods a bit further to establish the following generalization of Bang's theorem, published by Zsigmondy in 1892: *Let a, b be positive coprime integers with $a > b$. For each $n \geq 1$, there is a prime number p which divides $a^n - b^n$ but which does not divide $a^k - b^k$ for any positive integer $k < n$, except in the following cases:*

- $n = 1$ and $a - b = 1$,
- $n = 2$ and $a + b$ is a power of 2,
- $n = 6$, $a = 2$ and $b = 1$.

A clean exposition can be found in [4].

References

1. Paul Pollack, *A generalization of the Hardy–Ramanujan inequality and applications*. J. Number Theory 210 (2020), 171–182.
2. Carl Pomerance and Simon Rubinstein-Salzedo, *Cyclotomic Coincidences*. Experimental Mathematics, https://doi.org/10.1080/10586458.2019.1660741.
3. Srinivasa Ramanujan, *Highly composite numbers*, Proc. London Math. Soc. 14 (1915), 347–400.
4. Moshe Roitman, *On Zsigmondy primes*. Proc. Amer. Math. Soc. 125 (1997), 1913–1919.
5. Nairi Sedrakian and John Steinig, *A particular case of Dirichlet's theorem on arithmetic progressions*. Enseign. Math. (2) 44 (1998), 3–7.

Solutions to Step #12

12.118 Observe that

$$|\Omega(mn) - 2\log\log N| = |(\Omega(m) - \log\log N) + (\Omega(n) - \log\log N)|$$
$$\leq |\Omega(m) - \log\log N| + |\Omega(n) - \log\log N|.$$

Hence, if $|\Omega(mn) - 2\log\log N| > \epsilon \log\log N$, either m or n belongs to

$$\mathcal{E} := \{k \leq N : |\Omega(k) - \log\log N| > \frac{1}{2}\epsilon \log\log N\}.$$

Recall from Problem 11.112 that $\mathcal{E}$ contains asymptotically 0% of the integers in $[1, N]$, as $N \to \infty$. So given $\delta > 0$, we may choose $N_0(\delta)$ such that $\#\mathcal{E} < \delta N$ whenever $N > N_0(\delta)$. Then for $N > N_0(\delta)$, the number of pairs $(m, n) \in [1, N]^2$ with $m \in \mathcal{E}$ or $n \in \mathcal{E}$ is smaller than $\delta N \cdot N + N \cdot \delta N = 2\delta \cdot N^2$, and so a random pair from $[1, N]^2$ has this property with probability $< 2\delta$. Since $\delta > 0$ is arbitrary, the result follows.

12.119 Notice that $\log\log(N^2) - \log\log N = \log 2$. Hence, if N is large and $|\Omega(n) - \log\log N| > \epsilon \log\log N$, then

$$|\Omega(n) - \log\log(N^2)| \geq |\Omega(n) - \log\log N| - \log 2$$

$$> \epsilon \log\log N - \log 2 > \frac{1}{2}\epsilon \log\log(N^2).$$

By Problem 11.112, the proportion of $n \in [1, N^2]$ with this property tends to 0, as $N \to \infty$.

12.120 We fix $\epsilon > 0$ and argue that $\frac{M(N)}{N^2} < 2\epsilon$ for all large enough N.

Suppose that a appears in the $N \times N$ multiplication table.

© Springer Nature Switzerland AG 2021
P. Pollack, A. Singha Roy, *Steps into Analytic Number Theory*, Problem Books in
Mathematics, https://doi.org/10.1007/978-3-030-65077-3_27

If $|\Omega(a) - \log\log N| > \frac{1}{2}\log\log N$, then a belongs to a certain set of size $< \epsilon N^2$ by Problem 12.119 (once N is large enough). Suppose instead that $|\Omega(a) - \log\log N| \leq \frac{1}{2}\log\log N$. By assumption, we can write $a = mn$ with $(m, n) \in [1, N]^2$. Then

$$|\Omega(mn) - 2\log\log N| = |\Omega(a) - 2\log\log N| \geq \frac{1}{2}\log\log N.$$

By Problem 12.118, there are fewer than ϵN^2 possibilities for the pair (m, n) (once N is large enough), and so also fewer than ϵN^2 possibilities for the product mn.

12.121

(a) By Problem 8.82,

$$(2n)! = \prod_{m \leq 2n} \text{LCM}(2n/m)$$

and

$$n! = \prod_{m \leq n} \text{LCM}(n/m) = \prod_{\substack{m \leq 2n \\ m \text{ even}}} \text{LCM}(2n/m).$$

Hence,

$$\binom{2n}{n} = \frac{(2n)!}{(n!)^2} = \frac{\prod_{m \leq 2n} \text{LCM}(2n/m)}{\left(\prod_{\substack{m \leq 2n \\ m \text{ even}}} \text{LCM}(2n/m)\right)\left(\prod_{m \leq n} \text{LCM}(n/m)\right)}$$

$$= \frac{\prod_{\substack{m \leq 2n \\ m \text{ odd}}} \text{LCM}(2n/m)}{\prod_{m \leq n} \text{LCM}(n/m)};$$

this is the claimed representation of $\binom{2n}{n}$. The inequality follows from the unimodality of the binomial coefficients: Indeed, since $\binom{2n-1}{n-1} = \max_{0 \leq r \leq 2n-1}\binom{2n-1}{r}$,

$$\binom{2n}{n} = 2\binom{2n-1}{n-1} \geq \frac{2}{2n}\sum_{r=0}^{2n-1}\binom{2n-1}{r} = \frac{2 \cdot 2^{2n-1}}{2n} = \frac{4^n}{2n}.$$

(b) For all pairs of positive integers m, n, we have that $\frac{2n}{2m+1} \leq \frac{2n}{2m} = \frac{n}{m}$, and so $\text{LCM}(\frac{2n}{2m+1}) \leq \text{LCM}(\frac{n}{m})$. Thus, from part (a),

$$\frac{4^n}{2n} \le \binom{2n}{n} = \frac{\text{LCM}(2n)}{\text{LCM}(n)}\text{LCM}(2n/3) \cdot \prod_{m \ge 2} \frac{\text{LCM}(\frac{2n}{2m+1})}{\text{LCM}(\frac{n}{m})}$$

$$\le \frac{\text{LCM}(2n)}{\text{LCM}(n)}\text{LCM}(2n/3).$$

(c) By Problem 7.68, $\text{LCM}(2n/3) = \text{LCM}(\lfloor 2n/3 \rfloor) \le 4^{\lfloor 2n/3 \rfloor} \le 4^{2n/3}$. That $\frac{\text{LCM}(2n)}{\text{LCM}(n)} > \frac{4^{n/3}}{2n}$ is now immediate from (b).

(d) We argue as in Problem 7.65. Let p be a prime dividing $\text{LCM}(2n)$, and let p^k be the highest power of p dividing $\text{LCM}(2n)$. Then $p^k \le 2n$ (indeed, p^k is the highest power of p not exceeding $2n$). Thus, $p^{k-1} = p^k/p \le p^k/2 \le 2n/2 = n$. Hence, $p^{k-1} \mid \text{LCM}(n)$, and so $p^2 \nmid \frac{\text{LCM}(2n)}{\text{LCM}(n)}$. Thus, $\frac{\text{LCM}(2n)}{\text{LCM}(n)}$ is squarefree.

To finish off, it is enough to show that no prime $p \in (\sqrt{2n}, n]$ divides $\frac{\text{LCM}(2n)}{\text{LCM}(n)}$. For those primes p, the power p^1 is the highest power of p not exceeding n and also the highest power of p not exceeding $2n$. Hence, p appears in the factorizations of both $\text{LCM}(2n)$ and $\text{LCM}(n)$ to the first power, and so $p \nmid \frac{\text{LCM}(2n)}{\text{LCM}(n)}$.

(e) Suppose by way of contradiction that there is no prime in the interval $(n, 2n]$ for some $n \in \mathbb{Z}^+$. Then by parts (c) and (d),

$$\frac{4^{n/3}}{2n} \le \frac{\text{LCM}(2n)}{\text{LCM}(n)} \le \prod_{p \le \sqrt{2n}} p \le \text{LCM}(\sqrt{2n}) \le 4^{\lfloor \sqrt{2n} \rfloor} \le 4^{\sqrt{2n}}.$$

However, the inequality $\frac{4^{n/3}}{2n} \le 4^{\sqrt{2n}}$ fails once $n \ge 34$ (see below). Thus, as long as $n \ge 34$, there is a prime in $(n, 2n]$. Finally, for each $n < 34$, one of the primes $2, 3, 5, 7, 13, 23, 43$ belongs to $(n, 2n]$.

How can we see that the inequality fails for $n \ge 34$? Let $g(x) = \frac{4^{x/3}}{2x \cdot 4^{\sqrt{2x}}}$. It is enough to prove that $g(x) > 1$ for all real $x \ge 34$. For $x \ge 10$,

$$(\log g(x))' = \frac{\log 4}{3} - \frac{1}{x} - \frac{\log 4}{\sqrt{2x}} \ge \frac{\log 4}{3} - \frac{1}{10} - \frac{\log 4}{\sqrt{2 \cdot 10}} > 0,$$

and so g is increasing on $[10, \infty)$. It remains only to note that $g(34) = 1.06199\ldots > 1$.

12.122 When $n \ne 6$, there is a prime ℓ for which $o(2 \bmod \ell) = n$ (Exercise 11.117). Then $n \mid \ell - 1$, so that $\ell \equiv 1 \pmod{n}$, while $\ell \mid 2^n - 1$, so that $\ell < 2^n$. When $n = 6$, it is enough to observe that $\ell = 7$ is both 1 mod 6 and smaller than 2^6.

Remark. The following deep result on the least prime in a general coprime progression is due to Linnik: *For certain positive constants c_1 and c_2, and every pair of coprime integers a and m, with $m > 0$, there is a prime $p \equiv a \pmod{m}$ with $p \le c_1 m^{c_2}$.* That is, the least prime in a coprime progression is polynomially bounded by the size of the modulus. Xylouris has shown that c_2 can be taken as 5, for a suitable choice of c_1.

12.123 We deduce this from a lemma of independent interest: *For each fixed $\epsilon > 0$, and all large enough x, the interval $(x, (1+\epsilon)x]$ contains a prime.* (For $\epsilon = 1$, it follows from Exercise 12.121 that all $x \geq 1$ are "large enough", but in this problem we will need the lemma for arbitrarily small values of ϵ.)

To prove the lemma, it suffices to show that

$$\lim_{x \to \infty} \frac{\pi((1+\epsilon)x)}{\pi(x)} = 1 + \epsilon.$$

Rewrite

$$\frac{\pi((1+\epsilon)x)}{\pi(x)} = \frac{\pi((1+\epsilon)x)}{(1+\epsilon)x/\log((1+\epsilon)x)} \cdot \frac{(1+\epsilon)x/\log((1+\epsilon)x)}{x/\log x} \cdot \frac{x/\log x}{\pi(x)}.$$

By the Prime Number Theorem, the first and third right-hand factors tend to 1, as $x \to \infty$. The middle factor is

$$(1+\epsilon)\frac{\log x}{\log((1+\epsilon)x)} = (1+\epsilon)\frac{1}{1 + \frac{\log(1+\epsilon)}{\log x}},$$

which tends to $1 + \epsilon$.

Now let $A := \sum_{i=1}^{n} 10^{n-i} d_i$ be the (fixed) integer whose digits, from left-to-right, are $d_1, \ldots, d_n$. For each positive integer k, the number of primes whose decimal expansion begins with A and which have exactly $n + k$ digits is $\pi((A+1) \cdot 10^k) - \pi(A \cdot 10^k)$. This is positive for all large enough k, by the lemma with $\epsilon = \frac{1}{A}$.

12.124

(a) We apply Abel summation. Let $a_n = 1$ when n is prime and $a_n = 0$ otherwise, and take $f(x) = x$. Then with $S(x)$ having its usual meaning, $S(x) = \pi(x)$ and $f'(x) = 1$. Hence,

$$\sum_{p \leq x} p = \sum_{n \leq x} a_n f(n) = S(x)f(x) - \int_1^x S(t)f'(t)\,dt$$

$$= x \cdot \pi(x) - \int_2^x \pi(t)\,dt.$$

(We start the final integral at $x = 2$, since $\pi(x)$ vanishes when $x < 2$.) By the Prime Number Theorem, $\lim_{x \to \infty} \frac{x \cdot \pi(x)}{x^2/\log x} = \lim_{x \to \infty} \frac{\pi(x)}{x/\log x} = 1$. Thus, what needs to be shown is that

$$\lim_{x \to \infty} \frac{\int_2^x \pi(t)\,dt}{x^2/\log x} = \frac{1}{2}.$$

Define $E(t)$, for real numbers $t \geq 2$, by the equation $\pi(t) = \frac{t}{\log t} + E(t)$. By L'Hôpital's rule,

$$\lim_{x \to \infty} \frac{\int_2^x t/\log t \, dt}{x^2/\log x} = \lim_{x \to \infty} \frac{x/\log x}{(2x/\log x)(1 - \frac{1}{2}/\log x)} = \frac{1}{2}. \qquad (12.41)$$

Thus, all we have left to do is to show that $\lim_{x \to \infty} \frac{\int_2^x E(t) \, dt}{x/(\log x)^2} = 0$. We will prove the stronger assertion that

$$\lim_{x \to \infty} \frac{\int_2^x |E(t)| \, dt}{x^2/\log x} = 0. \qquad (12.42)$$

The Prime Number Theorem says that $\frac{E(t)}{t/\log t} \to 0$, as $t \to \infty$. So given $\epsilon > 0$, we have $|E(t)| < \epsilon \frac{t}{\log t}$ for all sufficiently large t. It follows that is a positive constant C (depending on ϵ) with $|E(t)| \leq C + \epsilon \frac{t}{\log t}$ for all $t \geq 2$. Keeping (12.41) in mind, we deduce that for all large enough x,

$$\frac{\int_2^x |E(t)| \, dt}{x^2/\log x} \leq \frac{\int_2^x (C + \epsilon t/\log t) \, dt}{x^2/\log x} \leq \frac{Cx}{x^2/\log x} + \epsilon \frac{\int_2^x t/\log t \, dt}{x^2/\log x} \leq \epsilon.$$

Since this holds for each $\epsilon > 0$, (12.42) follows.

(b) Let $y = x/(\log x)^2$ and decompose $A(n) = A_0(n) + A_1(n)$, where $A_0(n) = \sum_{p|n, \, p \leq y} p$, and $A_1(n) = \sum_{p|n, \, p > y} p$. Then

$$\sum_{n \leq x} A_0(n) = \sum_{n \leq x} \sum_{\substack{p|n \\ p \leq y}} p = \sum_{p \leq y} p \sum_{\substack{n \leq x \\ p|n}} 1 \leq \sum_{p \leq y} p \cdot \frac{x}{p} \leq xy = x^2/(\log x)^2,$$

so that

$$\lim_{x \to \infty} \frac{\sum_{n \leq x} A_0(n)}{x^2/\log x} = 0.$$

So the problem reduces to computing $\lim_{x \to \infty} \frac{\sum_{n \leq x} A_1(n)}{x^2/\log x}$.

We rewrite

$$\sum_{n \leq x} A_1(n) = \sum_{n \leq x} \sum_{\substack{p|n \\ p > y}} p = \sum_{\substack{m,p \\ mp \leq x \\ p > y}} p = \sum_{m \leq x/y} \sum_{y < p \leq x/m} p.$$

Clearly,

$$\sum_{y < p \le x/m} p \ge \left(\sum_{p \le x/m} p \right) - y^2.$$

Now fix $\epsilon \in (0, 1)$. For $m \le x/y$, we have $x/m \ge y = x/(\log x)^2$. Since $x/(\log x)^2 \to \infty$ as $x \to \infty$, we infer from part (a) that for all large x, and all $m \le x/y$,

$$\sum_{p \le x/m} p \ge (1 - \epsilon) \cdot \frac{1}{2}(x/m)^2 / \log(x/m) \ge (1 - \epsilon) \cdot \frac{1}{2m^2} x^2 / \log x.$$

Therefore,

$$\sum_{m \le x/y} \sum_{y < p \le x/m} p \ge \sum_{m \le x/y} \left((1 - \epsilon) \frac{1}{2m^2} x^2 / \log x - y^2 \right)$$

$$\ge \left((1 - \epsilon) \frac{x^2}{\log x} \sum_{m \le x/y} \frac{1}{2m^2} \right) - xy.$$

Dividing by $x^2/\log x$, we find that for all large enough x,

$$\frac{\sum_{n \le x} A_1(n)}{x^2/\log x} \ge (1 - \epsilon) \sum_{m \le x/y} \frac{1}{2m^2} - \frac{xy}{x^2/\log x}$$

$$\ge (1 - \epsilon) \left(\frac{1}{2} \zeta(2) - \epsilon \right) - \epsilon$$

$$\ge (1 - \epsilon) \cdot \frac{1}{2} \zeta(2) - 2\epsilon \ge \frac{1}{2} \zeta(2) - 3\epsilon.$$

We can get a nearly equal upper bound by a similar argument. This time, we use part (a) to deduce that for each fixed $\epsilon \in (0, 1)$, all large x, and all $m \le x/y$,

$$\sum_{y < p \le x/m} p \le \sum_{p \le x/m} p \le (1 + \epsilon) \cdot \frac{1}{2} \frac{(x/m)^2}{\log(x/m)}$$

$$\le (1 + \epsilon)^2 \cdot \frac{1}{2} \frac{(x/m)^2}{\log x}.$$

To bound $\log(x/m)$ from below we used that when x is large and $m \le x/y$,

$$\log(x/m) \ge \log y = \log x - 2 \log \log x > \frac{1}{1 + \epsilon} \log x.$$

Thus,

$$\sum_{m \le x/y} \sum_{p \le x/m} p \le (1+\epsilon)^2 \cdot \frac{x^2}{\log x} \sum_{m \le x/y} \frac{1}{2m^2}.$$

Now divide by $x^2/\log x$ to find that

$$\frac{\sum_{n \le x} A_1(n)}{x^2/\log x} \le (1+\epsilon)^2 \cdot \frac{1}{2}\zeta(2) \le (1+3\epsilon) \cdot \frac{1}{2}\zeta(2) \le \frac{1}{2}\zeta(2) + 3\epsilon.$$

Since ϵ can be arbitrarily small, comparing our upper and lower bounds shows that the limit in question is $\frac{1}{2}\zeta(2)$ $(= \frac{\pi^2}{12})$.

Remark on the Prime Number Theorem We have not attempted to prove the Prime Number Theorem here. The simplest proofs use methods of complex analysis and rest on the study of $\zeta(s)$ as a function of a *complex* variable s (an innovation of Riemann), whereas our focus in these problem sets has been on real variable methods.

A remarkably short and elegant proof, depending only on basic results from complex analysis, was found by Donald J. Newman in 1980. For a masterful exposition of Newman's proof (as modified by Korevaar), see Don Zagier's 4-page (!) note [1].

12.125 To classify pairs of multisets $\{a_1, \ldots, a_k\}$, $\{b_1, \ldots, b_\ell\}$ satisfying

$$\prod_i S_{a_i}(n) = \prod_j S_{b_j}(n) \quad \text{for all } n \in \mathbb{Z}^+, \tag{12.43}$$

it is clearly enough to consider the case when no a_i is equal to a b_j. When this condition holds, we call the pair $\{a_1, \ldots, a_k\}$, $\{b_1, \ldots, b_\ell\}$ *reduced*. We will show that the only reduced pairs are $\{1, 1, \ldots, 1\}$ and $\{3, 3, \ldots, 3\}$, where there are twice as many 1s as 3s.[1] Thus (perhaps disappointingly), the only interesting identity of this kind is the one already noted in the problem statement.

If $A = \{a_1, \ldots, a_k\}$, $B = \{b_1, \ldots, b_\ell\}$ is a reduced pair, then taking $n = 2$ in (12.43) reveals that

$$\prod_i (2^{a_i} + 1) = \prod_j (2^{b_j} + 1). \tag{12.44}$$

We use (12.44) to argue that each $a_i, b_j \le 3$. Suppose otherwise, and let m be the maximum of the a_i and b_j. Without loss of generality, we can assume $a_1 = m > 3$. By Bang's theorem (Problem 11.117), there is a prime p with order $2a_1$. Then

$$p \mid 2^{2a_1} - 1 = (2^{a_1} - 1)(2^{a_1} + 1).$$

---

[1] By "pairs" here we mean "unordered pairs". So, for instance, $\{1, 1\}$, $\{3\}$ is the same pair as $\{3\}$, $\{1, 1\}$.

Moreover, $p \nmid 2^{a_1} - 1$, as otherwise the order of 2 mod p would be at most a_1, and not equal to $2a_1$. Hence, p divides $2^{a_1} + 1$. It follows from (12.44) that $p \mid 2^{b_j} + 1$ for some j. But then p also divides $2^{2b_j} - 1$, so that the order of 2 modulo p is at most $2b_j$. Hence, $2a_1 \leq 2b_j$, and $a_1 \leq b_j$. Since A, B is reduced, the inequality is strict. But this contradicts that $a_1 = m$.

At this point we know that each $a_i, b_j \in \{1, 2, 3\}$. Suppose that some a_i or b_j is equal to 2; without loss of generality, say $a_1 = 2$. Then 5 divides the left hand side of (12.44) and so also divides the right. But the only positive integer $b \leq 3$ for which $5 \mid 2^b + 1$ is $b = 2$, and so some $b_j = 2$, contradicting that A, B is reduced. So each a_i, b_j is either 1 or 3. As A, B have no elements in common, we can assume A consists of k 1s and B consists of ℓ 3s. Then referring back to (12.44), $3^k = 9^\ell$, and so $k = 2\ell$.

Reference

1. Don Zagier, *Newman's short proof of the prime number theorem*. Amer. Math. Monthly 104 (1997), 705–708.

Solutions to Special Step A

In the remaining solution sets, we will sometimes refer to "standard" results from mathematical analysis. As references for this material, we recommend [1] (for real analysis) and [2] (for complex analysis).

13.126 Every element of $\mathbb{U}_8$ squares to the identity. (This amounts to the fact that the square of an odd number is always 1 mod 8.) Thus, if χ is any character of $\mathbb{U}_8$, and $g \in \mathbb{U}_8$, then

$$1 = \chi(1) = \chi(g^2) = \chi(g)^2,$$

so that $\chi(g) = \pm 1$.

To pin things down further, observe that each $g \in \mathbb{U}_8$ admits a representation $g = (3 \bmod 8)^a \cdot (5 \bmod 8)^b$ for some integers a, b. This representation is certainly not unique, but for a given g, the parities of a and b are uniquely determined. It follows that every function $\chi : \{3 \bmod 8, 5 \bmod 8\} \to \{-1, 1\}$ extends to a unique character of $\mathbb{U}_8$, by defining

$$\chi((3 \bmod 8)^a (5 \bmod 8)^b) = \chi(3 \bmod 8)^a \cdot \chi(5 \bmod 8)^b.$$

A little checking shows that this description of the characters χ matches the functions listed in the table.

13.127 In the row corresponding to 1 mod 8, every entry is 1, and the entries sum to 4. By direct computation, the row sum in every other row is 0. Therefore, the sum $\chi_1(g) + \chi_2(g) + \chi_3(g) + \chi_4(g)$ is equal to 4 when $g = 1 \bmod 8$, and equal to 0 otherwise. So by definition of the indicator function, $\mathcal{I}_{1 \bmod 8} = \frac{1}{4}(\chi_1 + \chi_2 + \chi_3 + \chi_4)$.

13.128 For every pair of odd a and b,

$$a \bmod 8 = b \bmod 8 \iff (a \bmod 8)^{-1} \cdot (b \bmod 8) = 1 \bmod 8.$$

© Springer Nature Switzerland AG 2021
P. Pollack, A. Singha Roy, *Steps into Analytic Number Theory*, Problem Books in Mathematics, https://doi.org/10.1007/978-3-030-65077-3_28

So by the previous problem,

$$\mathcal{I}_{a \bmod 8}(b \bmod 8) = \frac{1}{4} \sum_{\chi} \chi((a \bmod 8)^{-1}(b \bmod 8))$$

$$= \frac{1}{4} \sum_{\chi} \chi(a \bmod 8)^{-1} \chi(b \bmod 8).$$

This proves the first claim of the problem. The concrete expressions for $\mathcal{I}_{a \bmod 8}$ follow from looking at the table itself.

13.129 For all complex numbers s with $\Re(s) > 1$, we can write $s = \sigma + i\tau$ where $\sigma, \tau \in \mathbb{R}$ and $\sigma > 1$. Then for each positive integer n,

$$|n^s| = |n^{\sigma}| \cdot |n^{i\tau}| = n^{\sigma} \cdot |\cos(\tau \log n) + i \sin(\tau \log n)| = n^{\sigma}.$$

Hence, $\left|\frac{\chi(n)}{n^s}\right| \leq \frac{1}{n^{\sigma}}$ for each n (with equality if n and m are coprime). The convergence of $\sum_{n=1}^{\infty} \left|\frac{\chi(n)}{n^s}\right|$ now follows by comparison with $\sum_{n=1}^{\infty} \frac{1}{n^{\sigma}}$.

To get the product decomposition, we argue as in the solution to Problem 3.22 (or apply directly the remark at the end of the solution to Problem 3.25).

13.130 If we replace each term of the double sum with its absolute value, we arrive at

$$\sum_{p} \sum_{k \geq 1} \frac{|\chi(p^k)|}{kp^{ks}}.$$

This is dominated, term by term, by the double sum

$$\sum_{p} \sum_{k \geq 1} \frac{1}{kp^{ks}},$$

which was shown to converge in Problem 3.23 (in fact, to converge to $\log \zeta(s)$). So we have absolute convergence.

Recall that $\log \frac{1}{1-x} = x + \frac{1}{2}x^2 + \frac{1}{3}x^3 + \ldots$ for all real x with $|x| < 1$. When $s > 1$, the absolute value $\left|\frac{\chi(p)}{p^s}\right| \leq \frac{1}{p^s} < 1$ for all primes p. Hence,

$$\sum_{p} \sum_{k \geq 1} \frac{\chi(p^k)}{kp^{ks}} = \sum_{p} \log\left(\frac{1}{1 - \frac{\chi(p)}{p^s}}\right).$$

Exponentiating yields

$$\exp\left(\sum_p \sum_{k\geq 1} \frac{\chi(p^k)}{kp^{ks}}\right) = \prod_p \frac{1}{1 - \frac{\chi(p)}{p^s}} = L(s, \chi). \tag{A.45}$$

Since $L(s, \chi)$ can be written as the exponential of a real number, $L(s, \chi) > 0$. The expression claimed for $\log L(s, \chi)$ is now immediate from (A.45).

13.131

(a) This is a direct application of Abel summation, with each $a_n = \chi(n)$ and $f(t) = t^{-s}$.

(b) Examining the table, we find that $\sum_{g\in \mathbb{U}_8} \chi(g)$ vanishes for every nontrivial character χ of $\mathbb{U}_8$. As a consequence, for every nontrivial Dirichlet character χ mod 8, the sum of $\chi(n)$ over any 8 consecutive integers vanishes. Therefore, it is enough to verify that $|A_\chi(t)| \leq 2$ for $t \leq 7$. This is easy to check directly from the values in the table.

(c) The bound on $|A_\chi(t)|$ proved in (b) implies that the improper integral $\int_1^\infty \frac{A_\chi(t)}{t^{s+1}}\,dt$ converges absolutely, and that (for any $s > 0$) $A_\chi(x)x^{-s} \to 0$ as $x \to \infty$. Now sending x to infinity in the result of (a), we get that the series defining $L(s, \chi)$ converges to $s \int_1^\infty \frac{A_\chi(t)}{t^{s+1}}\,dt$.

(d) We may deduce from (a)–(c) that for all $x \geq 1$ and all $s \geq \epsilon$,

$$\left| L(s, \chi) - \sum_{n\leq x} \frac{\chi(n)}{n^s} \right| = \left| s \int_x^\infty \frac{A_\chi(t)}{t^{s+1}}\,dt - x^{-s} A_\chi(x) \right|$$

$$\leq 2s \int_x^\infty \frac{dt}{t^{s+1}} + 2x^{-s} = 4x^{-s} \leq 4x^{-\epsilon}.$$

Given any $\delta > 0$, we have $4x^{-\epsilon} < \delta$ whenever $x > (4/\delta)^{1/\epsilon}$. That bound on x is independent of s, proving uniform convergence for $s \geq \epsilon$.

Since each summand $\chi(n)n^{-s}$ is a continuous function of s, continuity of the sum function $L(s, \chi)$ on $s \geq \epsilon$ now follows from standard theorems. As ϵ can be any positive number, $L(s, \chi)$ is continuous for all $s > 0$.

(e) By part (d), $L(s, \chi)$ is continuous at $s = 1$. Since $L(s, \chi) > 0$ for $s > 0$, continuity guarantees that $L(1, \chi) \geq 0$. We are supposing that $L(1, \chi) \neq 0$, and so $L(1, \chi) > 0$.

Now consider the composite function $\log L(s, \chi)$. Since $L(s, \chi)$ is continuous and positive-valued for $s \geq 1$, this composite function is continuous for $s \geq 1$. Being continuous, $\log L(s, \chi)$ is certainly bounded on the closed interval $[1, 2]$. That is, $\log L(s, \chi) = O(1)$ for $1 \leq s \leq 2$.

13.132

(a) For $s > 1$, we have from Problem 13.130 and our solution to Problem 3.23 that

$$\left| \sum_p \frac{\chi(p)}{p^s} - \log L(s, \chi) \right| = \left| \sum_{\substack{p,k \\ k \geq 2}} \frac{\chi(p)}{kp^{ks}} \right| \leq \sum_{\substack{p,k \\ k \geq 2}} \frac{1}{kp^{ks}} < 1.$$

(b) By definition of χ_0,

$$L(s, \chi_0) = \sum_{n=1}^{\infty} \frac{\chi_0(n)}{n^s} = \sum_{n \text{ odd}} \frac{1}{n^s}$$

$$= \sum_{n \geq 1} \frac{1}{n^s} - \sum_{m \geq 1} \frac{1}{(2m)^s} = \left(1 - \frac{1}{2^s}\right) \zeta(s).$$

By part (a) and Problem 3.24, we obtain, for all $s \in (1, 2)$,

$$\sum_p \frac{\chi_0(p)}{p^s} = \log L(s, \chi_0) + O(1)$$

$$= \log \zeta(s) + \log\left(1 - \frac{1}{2^s}\right) + O(1) = \log \frac{1}{s-1} + O(1).$$

Here we used that $\log\left(1 - \frac{1}{2^s}\right) \in \left(\log \frac{1}{2}, \log \frac{3}{4}\right)$ when $1 < s < 2$.

(c) This is an obvious corollary of part (a) and Problem 13.131(e).

13.133 For every odd integer a, and every $s > 1$, the computation preceding Problem 13.132 tells us that

$$\sum_{p \equiv a \pmod 8} \frac{1}{p^s} = \frac{1}{4} \sum_p \frac{\chi_0(p)}{p^s} + \frac{1}{4} \sum_{\chi \neq \chi_0} \chi(a)^{-1} \sum_p \frac{\chi(p)}{p^s}.$$

Substituting in the estimates from the last problem,

$$\sum_{p \equiv a \pmod 8} \frac{1}{p^s} = \frac{1}{4} \log \frac{1}{s-1} + O(1)$$

for $1 < s < 2$, as desired. Now letting $s \downarrow 1$ yields the divergence of the series $\sum_{p \equiv a \pmod 8} \frac{1}{p}$ and, consequently, the infinitude of primes $p \equiv a \pmod 8$.

13.134 For χ_1, the grouping of terms

$$L(1, \chi_1) = \left(1 - \frac{1}{3}\right) + \left(\frac{1}{5} - \frac{1}{7}\right) + \left(\frac{1}{9} - \frac{1}{11}\right) + \left(\frac{1}{13} - \frac{1}{15}\right)$$

makes it obvious that $L(1, \chi_1) > 0$. To handle χ_3, we rewrite

$$L(1, \chi_3) = \left(1 + \frac{1}{3}\right) - \left(\frac{1}{5} + \frac{1}{7}\right) + \left(\frac{1}{9} + \frac{1}{11}\right) - \left(\frac{1}{13} + \frac{1}{15}\right) + \cdots.$$

This is an alternating series with terms strictly decreasing in absolute value; hence, $L(1, \chi_3) > (1 + \frac{1}{3}) - (\frac{1}{5} + \frac{1}{7}) > 0$. Finally,

$$L(1, \chi_2) = 1 - \left(\frac{1}{3} + \frac{1}{5}\right) + \left(\frac{1}{7} + \frac{1}{9}\right) - \left(\frac{1}{11} + \frac{1}{13}\right) + \cdots,$$

so that

$$1 - L(1, \chi_2) = \left(\frac{1}{3} + \frac{1}{5}\right) - \left(\frac{1}{7} + \frac{1}{9}\right) + \cdots.$$

The right-hand side is an alternating series with terms strictly decreasing in absolute value, and so is $< \frac{1}{3} + \frac{1}{5} < 1$. Thus, $L(1, \chi_2) > 0$.

13.135 The groups $\mathbb{U}_8$ and $\mathbb{U}_{12}$ are isomorphic, both being abstractly the same as $\mathbb{Z}_2 \oplus \mathbb{Z}_2$. An explicit isomorphism maps the two independent generators 3 mod 8 and 5 mod 8 of $\mathbb{U}_8$ to the two independent generators 5 mod 12 and 7 mod 12 of $\mathbb{U}_{12}$. From this, we infer that $\mathbb{U}_{12}$ has the "same" character table as $\mathbb{U}_8$, if we relabel the row headers with 1 mod 12, 5 mod 12, 7 mod 12, and 11 mod 12:

	χ_0	χ_1	χ_2	χ_3
1 mod 12	1	1	1	1
5 mod 12	1	−1	−1	1
7 mod 12	1	1	−1	−1
11 mod 12	1	−1	1	−1

All of our subsequent arguments go through with only minor changes. Perhaps the only difference worth noting is that the formula $L(s, \chi_0) = (1 - 2^{-s})\zeta(s)$, from the start of Problem 13.132(b), is now replaced by the identity

$$L(s, \chi_0) = (1 - 2^{-s})(1 - 3^{-s})\zeta(s).$$

To prove this, notice that $\chi_0(2) = \chi_0(3) = 0$, while $\chi_0(p) = 1$ for all primes $p > 3$; therefore,

$$L(s, \chi_0) = \prod_p \frac{1}{1 - \frac{\chi_0(p)}{p^s}} = \prod_{p \neq 2,3} \frac{1}{1 - \frac{1}{p^s}}$$

$$= (1 - 2^{-s})(1 - 3^{-s}) \prod_p \frac{1}{1 - \frac{1}{p^s}} = (1 - 2^{-s})(1 - 3^{-s})\zeta(s).$$

Note that this change to the first half of Problem 13.132(b) does not affect the deduction that $\sum_p \frac{\chi_0(p)}{p^s} = \log \frac{1}{s-1} + O(1)$.

References

1. Stephen Abbott, *Understanding analysis*. Second edition. Undergraduate Texts in Mathematics. Springer, New York, 2015.
2. Joseph Bak and Donald J. Newman, *Complex analysis*. Third edition. Undergraduate Texts in Mathematics. Springer, New York, 2010.

Solutions to Special Step B

14.136 By Fermat's little theorem, $g^{\ell-1} = 1 \bmod \ell$ for all $g \in \mathbb{U}_\ell$. Thus, for all $g \in \mathbb{U}_\ell$,

$$\chi(g)^{\ell-1} = \chi(g^{\ell-1}) = \chi(1 \bmod \ell) = 1,$$

meaning that $\chi(g)$ is an $(\ell - 1)$th root of unity.

14.137 Let ω be a complex $(\ell - 1)$th root of unity. If there is any character χ of $\mathbb{U}_\ell$ with $\chi(g) = \omega$, then

$$\chi(g^k) = \omega^k \quad \text{for all } k \in \mathbb{Z}. \tag{B.46}$$

Since g generates $\mathbb{U}_\ell$, the description (B.46) completely determines χ. Thus, there can be at most one character χ of $\mathbb{U}_\ell$ with $\chi(g) = \omega$.

It remains to check that there is any such χ. For this, we simply define χ by (B.46). One might worry about whether χ is well-defined, but this is easily checked: If $g^{k_1} = g^{k_2}$, then $k_1 \equiv k_2 \pmod{\ell - 1}$, and so $\omega^{k_1} = \omega^{k_2}$. That χ is a homomorphism is immediate, and clearly $\chi(g) = \omega$.

Since there are $\ell - 1$ distinct $(\ell - 1)$th roots of unity, there are $\ell - 1$ characters of $\mathbb{U}_\ell$.

14.138 Let $a \in \mathbb{U}_\ell$ with $a \neq 1 \bmod \ell$.

We first argue that we can find a character ψ of $\mathbb{U}_\ell$ with $\psi(a) \neq 1$. To see this, fix a generator g of $\mathbb{U}_\ell$, and (using the previous problem) choose a character ψ of $\mathbb{U}_\ell$ with $\psi(g) = e^{2\pi i/(\ell-1)}$. Since a is not the identity of $\mathbb{U}_\ell$, $a = g^m$ for some m not divisible by $\ell - 1$, and

$$\psi(a) = \psi(g^m) = \psi(g)^m = e^{2\pi i \frac{m}{\ell-1}} \neq 1.$$

© Springer Nature Switzerland AG 2021
P. Pollack, A. Singha Roy, *Steps into Analytic Number Theory*, Problem Books in
Mathematics, https://doi.org/10.1007/978-3-030-65077-3_29

Now we show that the row sum vanishes in the row headed by a. Call that row sum S_a. Then

$$S_a = \sum_\chi \chi(a)$$

where χ runs over the characters of $\mathbb{U}_\ell$. Fix a character ψ of $\mathbb{U}_\ell$ with $\psi(a) \neq 1$. Then

$$\psi(a)S_a = \sum_\chi \psi(a)\chi(a).$$

For each character χ of $\mathbb{U}_\ell$, the function $\psi\chi$ on $\mathbb{U}_\ell$ defined by $(\psi\chi)(a) = \psi(a)\chi(a)$ is itself a character of $\mathbb{U}_\ell$. Moreover, distinct χ give rise to distinct functions $\psi\chi$. It follows that $\psi\chi$ runs over all the characters of $\mathbb{U}_\ell$ as χ does, and so

$$\sum_\chi \psi(a)\chi(a) = \sum_\chi (\psi\chi)(a) = \sum_\chi \chi(a) = S_a.$$

Comparing the last two displays, we find that S_a is invariant under multiplication by $\psi(a)$. Since $\psi(a) \neq 1$, this forces $S_a = 0$.

Thus, every row sum vanishes, except in the row headed by $1 \bmod \ell$. Of course, in that row every entry is 1, and so that row sum is $\ell - 1$.

From the above, we deduce that $\mathcal{I}_{1 \bmod \ell} = \frac{1}{\ell-1}\sum_\chi \chi$. The claimed identity for $\mathcal{I}_{a \bmod \ell}$ follows exactly as in Problem 13.128, since (for integers a and b coprime to ℓ)

$$a \bmod \ell = b \bmod \ell \iff (a \bmod \ell)^{-1}(b \bmod \ell) = 1 \bmod \ell.$$

14.139 Every compact subset of $\Re(s) > 1$ is contained within some half-plane $\Re(s) \geq 1 + \epsilon$, where $\epsilon > 0$. Thus, it is enough to show uniform convergence on each of these half-planes.

The terms in our double sum have the form $\epsilon_{p^k}\, p^{-ks}$, where $|\epsilon_{p^k}| \leq 1/k \leq 1$. When $\Re(s) \geq 1 + \epsilon$,

$$\left|\frac{\epsilon_{p^k}}{p^{ks}}\right| \leq \frac{1}{(p^k)^{1+\epsilon}}.$$

The sum $\sum_{p,k} \frac{1}{(p^k)^{1+\epsilon}}$ converges—and as a positive series, does so under an arbitrary ordering of its terms. So by the Weierstrass M-test, our series converges uniformly for $\Re(s) \geq 1 + \epsilon$, under any ordering of its terms.

14.140

(a) By definition of χ_0, the sum of the entries in the column headed by χ_0 is $\sum_{g \in \mathbb{U}_\ell} \chi_0(g) = |\mathbb{U}_\ell| = \ell - 1$. Now let χ be a nontrivial character of $\mathbb{U}_\ell$.

Choose $b \in \mathbb{U}_\ell$ with $\chi(b) \neq 1$. Writing S_χ for the sum of the entries in the column headed by χ,

$$\chi(b)S_\chi = \chi(b) \sum_{a \in \mathbb{U}_\ell} \chi(a) = \sum_{a \in \mathbb{U}_\ell} \chi(ba) = \sum_{a \in \mathbb{U}_\ell} \chi(a) = S_\chi.$$

Here we used that ba runs over the distinct elements of $\mathbb{U}_\ell$ as a does. Since $\chi(b) \neq 1$, it must be that $S_\chi = 0$.

(b) It follows from (a) that the sum of $\chi(n)$ vanishes when n runs over any block of ℓ consecutive integers. So it is enough to prove that $|\sum_{n \leq t} \chi(n)| < \ell$ when $t < \ell$. But this is immediate from the triangle inequality.

(c) It is enough to show uniform convergence on each set of the form

$$\{s \in \mathbb{C} : \Re(s) \geq \epsilon, |s| < 1/\epsilon\}, \tag{B.47}$$

where $\epsilon > 0$. We will prove this by borrowing some ideas from our solution to Problem 13.131. Throughout the next paragraph, we assume s belongs to the set (B.47).

Abel summation shows that for every real $x \geq 1$,

$$\sum_{n \leq x} \chi(n)n^{-s} = x^{-s}A_\chi(x) + s \int_1^x \frac{A_\chi(t)}{t^{s+1}} \, dt, \tag{B.48}$$

where $A_\chi(t) := \sum_{n \leq t} \chi(n)$. The improper integral $\int_1^\infty \frac{A_\chi(t)}{t^{s+1}} \, dt$ converges absolutely, since $|A_\chi(t)t^{-s-1}| \leq \ell t^{-1-\epsilon}$. Moreover, $|A_\chi(x)x^{-s}| \leq \ell x^{-\epsilon}$, and so $A_\chi(x)x^{-s} \to 0$ as $x \to \infty$. Thus, sending $x \to \infty$ in (B.48),

$$L(s, \chi) = s \int_1^\infty \frac{A_\chi(t)}{t^{s+1}} \, dt.$$

Moreover, writing σ for the real part of s, we have (again for $x \geq 1$)

$$\left| L(s, \chi) - \sum_{n \leq x} \frac{\chi(n)}{n^s} \right| = \left| s \int_x^\infty \frac{A_\chi(t)}{t^{s+1}} \, dt - x^{-s}A_\chi(s) \right|$$

$$\leq |s|\ell \int_x^\infty \frac{dt}{t^{\sigma+1}} + \ell x^{-\sigma}$$

$$= \ell \left(\frac{|s|}{\sigma} + 1 \right) x^{-\sigma}$$

$$\leq \ell \left(\frac{1}{\epsilon^2} + 1 \right) x^{-\epsilon}.$$

This final expression can be made arbitrarily small by choosing x large enough in terms of ϵ (independently of the precise value of s). So the series defining $L(s, \chi)$ converges uniformly on the region (B.47).

Finally, since each summand $\chi(n)n^{-s}$ is an entire function of s, standard results imply that $L(s, \chi)$ is holomorphic for $\Re(s) > 0$.

14.141 As mentioned in the problem statement, there is a logarithm $F(s)$ of $L(s, \chi)$ defined and analytic on open ball U centered at $s = 1$. On the other hand, by Exercise 14.139, the function $\operatorname{Log} L(s, \chi)$ itself is analytic for $\Re(s) > 1$.

Let $V = U \cap \{s \in \mathbb{C} : \Re(s) > 1\}$. Then V is a complex domain (a nonempty connected open set). Since $F(s)$ and $\operatorname{Log} L(s, \chi)$ are both logarithms of $L(s, \chi)$ on V, their difference $F(s) - \operatorname{Log} L(s, \chi)$ takes all its values on V from the set $\{2\pi i k : k \in \mathbb{Z}\}$. Since $F(s) - \operatorname{Log} L(s, \chi)$ is analytic on V (and in particular, continuous), there is a single integer k with $F(s) - \operatorname{Log} L(s, \chi) = 2\pi i k$ for all $s \in V$. Replacing $F(s)$ by $F(s) - 2\pi i k$, we can assume that $F(s) = \operatorname{Log} L(s, \chi)$ on V. We now extend $\operatorname{Log} L(s, \chi)$ to $U \cup \{s \in \mathbb{C} : \Re(s) > 1\}$ by defining $\operatorname{Log} L(s, \chi)$ to equal $F(s)$ when $s \in U \setminus \{s \in \mathbb{C} : \Re(s) > 1\}$. This provides an analytic continuation of $\operatorname{Log} L(s, \chi)$ to U.

The (continued version of the) function $\operatorname{Log} L(s, \chi)$, being analytic in a neighborhood of 1, is bounded on each sufficiently small closed ball centered at 1. So if we fix δ positive but sufficiently small, then $\operatorname{Log} L(s, \chi) = O(1)$ for all $s \in [1, 1 + \delta]$. On the other hand, by Problem 14.139, $\operatorname{Log} L(s, \chi)$ is analytic for $\Re(s) > 1$, and so in particular is continuous for real $s > 1$. Therefore, $\operatorname{Log} L(s, \chi)$ is bounded on the closed interval $[1 + \delta, 2]$. Putting these facts together, $\operatorname{Log} L(s, \chi) = O(1)$ for $1 \leq s \leq 2$. Note in these arguments, the implied constants may depend on χ.

14.142 This is true even without the assumption that ℓ is prime: For $1 < s < 2$, Problem 3.24 gives us that

$$\sum_p \frac{\chi_0(p)}{p^s} = \sum_{p:\ \gcd(p,\ell)=1} \frac{1}{p^s} = \sum_p \frac{1}{p^s} - \sum_{p|\ell} \frac{1}{p^s} = \log \frac{1}{s-1} + O_\ell(1).$$

In the final step, we used that $0 \leq \sum_{p|\ell} \frac{1}{p^s} \leq \sum_{p|\ell} \frac{1}{p}$, so that $\sum_{p|\ell} \frac{1}{p^s} = O_\ell(1)$.

14.143

(a) The proof is the same as that of Problem 13.132(a).

(b) This is an immediate consequence of part (a) and Problem 14.141.

14.144 Let a be an integer not divisible by ℓ. By Problems 14.142, 14.143, and the computation preceding Problem 14.142, when $1 < s < 2$,

$$\sum_{p \equiv a \ (\mathrm{mod}\ \ell)} \frac{1}{p^s} = \frac{1}{\ell-1} \sum_p \frac{\chi_0(p)}{p^s} + \frac{1}{\ell-1} \sum_{\chi \neq \chi_0} \chi(a)^{-1} \sum_p \frac{\chi(p)}{p^s}$$

$$= \frac{1}{\ell-1} \log \frac{1}{s-1} + O_\ell(1).$$

Letting $s \downarrow 1$, we deduce that there are infinitely many primes $p \equiv a \pmod{\ell}$.

14.145

(a) Let χ be either χ_1 or χ_2. Since $L(s, \chi)$ is analytic at $s = 1$ and $L(1, \chi) = 0$,

$$\lim_{s \downarrow 1} \frac{L(s, \chi)}{s - 1} = \lim_{s \downarrow 1} \frac{L(s, \chi) - L(1, \chi)}{s - 1} = L'(1, \chi).$$

Thus, $\dfrac{L(s, \chi_1)}{s - 1}$ and $\dfrac{L(s, \chi_2)}{s - 1}$ remain bounded as $s \downarrow 1$.
For $s > 1$,

$$L(s, \chi_0) = \prod_p \frac{1}{1 - \frac{\chi_0(p)}{p^s}} = \prod_{p:\, \gcd(p, \ell) = 1} \frac{1}{1 - \frac{1}{p^s}}$$

$$= (1 - \ell^{-s}) \prod_p \frac{1}{1 - \frac{1}{p^s}} = (1 - \ell^{-s}) \zeta(s).$$

Since $\zeta(s)(s - 1) \to 1$ as $s \downarrow 1$ (see Problem 2.10), we infer that

$$(s - 1) L(s, \chi_0) \to 1 - \ell^{-1}$$

as $s \downarrow 1$. Hence, $(s - 1) L(s, \chi_0)$ remains bounded as $s \downarrow 1$.
To finish off, write

$$\frac{1}{s - 1} \prod_\chi L(s, \chi) = ((s - 1) L(s, \chi_0)) \cdot \frac{L(s, \chi_1)}{s - 1} \cdot \frac{L(s, \chi_2)}{s - 1} \cdot \prod_{\chi \neq \chi_0, \chi_1, \chi_2} L(s, \chi).$$

The right-hand side is a finite product of terms remaining bounded as $s \downarrow 1$, and so itself remains bounded as $s \downarrow 1$.

(b) Observe that when $s > 1$,

$$\sum_\chi \text{Log } L(s, \chi) = \sum_\chi \sum_{\substack{p \text{ prime} \\ k \geq 1}} \frac{\chi(p^k)}{k p^{ks}} = \sum_{\substack{p \text{ prime} \\ k \geq 1}} \frac{1}{k p^{ks}} \sum_\chi \chi(p^k)$$

$$= (\ell - 1) \sum_{\substack{p, k \\ p^k \equiv 1 \;(\text{mod } \ell)}} \frac{1}{k p^{ks}} \geq 0.$$

(Since there are only finitely many characters χ mod ℓ, we encounter no difficulties when interchanging the order of summation.) Exponentiating, $\prod_\chi L(s, \chi) \geq 1$. Thus, $\frac{1}{s-1} \prod_\chi L(s, \chi) \to \infty$ as $s \downarrow 1$, contrary to what we found in (a). This shows that there can be at most one nontrivial Dirichlet character χ modulo ℓ satisfying $L(1, \chi) = 0$.

14.146

(a) Suppose for the sake of contradiction that $L(1, \chi) = 0$ where $\chi(\mathbb{Z}) \not\subseteq \mathbb{R}$. Then the function $\overline{\chi}(n)$, defined by $\overline{\chi}(n) = \overline{\chi(n)}$ for all n, is a Dirichlet character modulo ℓ distinct from χ. Furthermore,

$$0 = \overline{L(1, \chi)} = \overline{\sum_{n=1}^{\infty} \frac{\chi(n)}{n}} = \sum_{n=1}^{\infty} \frac{\overline{\chi}(n)}{n} = L(1, \overline{\chi}).$$

This contradicts the conclusion of Problem 14.145(b).

(b) Suppose χ is a nontrivial real-valued Dirichlet character modulo ℓ. Let g be an integer whose reduction mod ℓ generates $\mathbb{U}_\ell$. Then $\chi(g)$ is a real number of modulus 1, so that $\chi(g) = \pm 1$. If $\chi(g) = 1$, then $\chi(a) = 1$ whenever a coprime to ℓ, forcing $\chi = \chi_0$, contrary to hypothesis. So $\chi(g) = -1$. But then on integers a coprime to ℓ,

$$\chi(a) = 1 \iff a \text{ is congruent, mod } \ell, \text{ to an even power of } g$$

$$\iff a \text{ is a square modulo } \ell.$$

This shows that $\chi(a) = \left(\frac{a}{\ell}\right)$ when $\gcd(a, \ell) = 1$. Equality also holds when $\ell \mid a$, since then $\chi(a)$ and $\left(\frac{a}{\ell}\right)$ both vanish. Hence, $\chi(\cdot) = \left(\frac{\cdot}{\ell}\right)$.

14.147 We preface the solution with a comment on notation. Throughout this solution, "log" denotes the "principal branch" of the complex logarithm, defined for nonzero complex z by

$$\log z = \log |z| + i \arg z,$$

where $\arg z$ is the argument of z belonging to $(-\pi, \pi]$. It is well-known that

$$-\log(1 - z) = z + \frac{z^2}{2} + \frac{z^3}{3} + \cdots$$

for all z with $|z| \le 1$, $z \ne 1$.[1]

Coming to the proof proper, recall that the Gauss sum G admits the expression

---

[1]For $|z| < 1$, this follows in a routine way from the theory of Taylor expansions of holomorphic functions. The cases when $|z| = 1$, $z \ne 1$ are more delicate. Here one must apply Abel's lemma (see, e.g., Theorem 21.0 on p. 368 of [1]). Specifically, suppose that $|z| = 1$, $z \ne 1$. The partial sums of z^n are easily seen to be bounded (e.g., by $2/|1 - z|$); this implies, via summation by parts, that $\sum_{n \ge 1} \frac{z^n}{n}$ converges. Abel's lemma now shows that the power series $\sum_{n \ge 1} \frac{z^n}{n} t^n$ represents a continuous function of t on $[0, 1]$, so that $\sum_{n \ge 1} \frac{z^n}{n} = \lim_{t \uparrow 1}(-\log(1 - tz)) = -\log(1 - z)$.

$$G = \sum_{b \bmod \ell} \left(\frac{b}{\ell}\right) e^{2\pi i b/\ell}.$$

(See Eq. (8.25) from the solution of Problem 8.86.) Thus,

$$G \cdot L = \left(\sum_{b \bmod \ell} \left(\frac{b}{\ell}\right) e^{2\pi i b/\ell}\right) \left(\sum_{n=1}^{\infty} \left(\frac{n}{\ell}\right) \frac{1}{n}\right)$$

$$= \sum_{n=1}^{\infty} \frac{1}{n} \sum_{b \bmod \ell} \left(\frac{b}{\ell}\right)\left(\frac{n}{\ell}\right) e^{2\pi i b/\ell}.$$

When $\ell \mid n$, the Legendre symbol $\left(\frac{n}{\ell}\right) = 0$. So there is no harm in removing the terms where $\ell \mid n$. Now writing $\bar{n}$ for the inverse of $n \bmod \ell$, and noting that $\left(\frac{\bar{n}}{\ell}\right) = \left(\frac{n}{\ell}\right)$, we see that

$$\sum_{n=1}^{\infty} \frac{1}{n} \sum_{b \bmod \ell} \left(\frac{b}{\ell}\right)\left(\frac{n}{\ell}\right) e^{2\pi i b/\ell} = \sum_{\substack{n \geq 1 \\ \ell \nmid n}} \frac{1}{n} \sum_{b \bmod \ell} \left(\frac{b\bar{n}}{\ell}\right) e^{2\pi i b/\ell}.$$

We make the change of variables $a \equiv b\bar{n}$, so that $b \equiv an \pmod{\ell}$. This shows that

$$\sum_{\substack{n \geq 1 \\ \ell \nmid n}} \frac{1}{n} \sum_{b \bmod \ell} \left(\frac{b\bar{n}}{\ell}\right) e^{2\pi i b/\ell} = \sum_{\substack{n \geq 1 \\ \ell \nmid n}} \frac{1}{n} \sum_{a \bmod \ell} \left(\frac{a}{\ell}\right) e^{2\pi i an/\ell}.$$

We now restore the terms where $\ell \mid n$, noting that $\sum_{a \bmod \ell} \left(\frac{a}{\ell}\right) e^{2\pi i an/\ell} = 0$ for these n. We find that

$$\sum_{\substack{n \geq 1 \\ \ell \nmid n}} \frac{1}{n} \sum_{a \bmod \ell} \left(\frac{a}{\ell}\right) e^{2\pi i an/\ell} = \sum_{n=1}^{\infty} \frac{1}{n} \sum_{a \bmod \ell} \left(\frac{a}{\ell}\right) e^{2\pi i an/\ell}$$

$$= \sum_{n=1}^{\infty} \frac{1}{n} \sum_{\substack{a \bmod \ell \\ a \neq 0}} \left(\frac{a}{\ell}\right) e^{2\pi i an/\ell} = \sum_{\substack{a \bmod \ell \\ a \neq 0}} \left(\frac{a}{\ell}\right) \sum_{n=1}^{\infty} \frac{1}{n} e^{2\pi i an/\ell}. \tag{B.49}$$

The inner sum in the last expression is precisely $-\log(1 - e^{2\pi i a/\ell})$. Collecting our results,

$$G \cdot L = -\sum_{0 < a < \ell} \left(\frac{a}{\ell}\right) \log(1 - e^{2\pi i a/\ell}). \tag{B.50}$$

Thus, $\exp(G \cdot L) = \displaystyle\prod_{0 < a < \ell} (1 - e^{2\pi i a/\ell})^{-\left(\frac{a}{\ell}\right)} = A$, as claimed.

Remark. There is a surprising amount of arithmetic content hiding in the identity (B.50). We illustrate this by deducing two theorems of Dirichlet concerning primes $\ell \equiv 3 \pmod 4$.

When $\ell \equiv 3 \pmod 4$, we showed in Problem 5.45 that $G = \pm i\sqrt{\ell}$. In fact, the correct sign to choose is always $+$, as discussed in the remark following that problem. Turning to L, we know[2] that $L(s, \left(\frac{\cdot}{\ell}\right)) > 0$ for all $s > 1$ and so by continuity $L = L(1, \left(\frac{\cdot}{\ell}\right)) \geq 0$. We will show shortly that $L \neq 0$, and so $L > 0$. It follows that the left-hand side of (B.50) is purely imaginary, with positive imaginary part. What about the right-hand side? It is straightforward to check (draw a picture!) that $\arg(1 - e^{2\pi i a/\ell}) = \frac{\pi a}{\ell} - \frac{\pi}{2}$ whenever $0 < a < \ell$. Thus, the RHS in (B.50) has imaginary component

$$-\sum_{0 < a < \ell} \left(\frac{a}{\ell}\right)\left(\frac{\pi a}{\ell} - \frac{\pi}{2}\right) = -\frac{\pi}{\ell} \sum_{0 < a < \ell} a\left(\frac{a}{\ell}\right).$$

Now equating imaginary parts leads to the conclusion that

$$\sum_{0 < a < \ell} a\left(\frac{a}{\ell}\right) < 0. \tag{B.51}$$

In other words, when ℓ is a prime congruent to 3 mod 4, the sum of the modular nonsquares in $(0, \ell)$ exceeds the sum of the modular squares from that same interval.

We can take this analysis a bit further. It is an elementary (but challenging) exercise to show that when $\ell \equiv 3 \pmod 4$,

$$\sum_{0 < a < \ell/2} \left(\frac{a}{\ell}\right) = -\frac{1}{\ell}\left(2 - \left(\frac{2}{\ell}\right)\right) \sum_{0 < a < \ell} a\left(\frac{a}{\ell}\right).$$

(See, for instance, Example 7 on p. 284 of [2].) So from (B.51),

$$\sum_{0 < a < \ell/2} \left(\frac{a}{\ell}\right) > 0.$$

Hence, when $\ell \equiv 3 \pmod 4$, there are more modular squares than nonsquares in $(0, \ell/2)$.

These phenomena are peculiar to primes $\ell \equiv 3 \pmod 4$. For primes $\ell \equiv 1 \pmod 4$, it is easy to prove that the sum of the modular squares in $(0, \ell)$ coincides with the sum of the nonsquares, and that there are the same number of modular squares as nonsquares in $(0, \ell/2)$.

14.148 By definition,

$$A = \frac{\prod_{0 < a < \ell,\ \left(\frac{a}{\ell}\right) = -1} \left(1 - e^{2\pi i a/\ell}\right)}{\prod_{0 < a < \ell,\ \left(\frac{a}{\ell}\right) = 1} \left(1 - e^{2\pi i a/\ell}\right)}.$$

Assuming for the sake of contradiction that $L = 0$, we infer from Problem 14.147 that $A = \exp(G \cdot L) = \exp(0) = 1$.

---

[2]For instance, because $\mathrm{Log}\, L(s, \left(\frac{\cdot}{\ell}\right))$ is defined and real-valued when $s > 1$.

(a) As r runs through a complete set of representatives of the nonzero squares mod ℓ, the integer nr runs through a complete set of representatives of the nonsquares mod ℓ. Thus, with $\omega = e^{2\pi i/\ell}$,

$$P(\omega) = \frac{\prod_{0<r<\ell,\,(\frac{r}{\ell})=1}(1-\omega^{nr})}{\prod_{0<r<\ell,\,(\frac{r}{\ell})=1}(1-\omega^{r})} = \frac{\prod_{0<a<\ell,\,(\frac{a}{\ell})=-1}\left(1-e^{2\pi ia/\ell}\right)}{\prod_{0<a<\ell,\,(\frac{a}{\ell})=1}\left(1-e^{2\pi ia/\ell}\right)} = A = 1,$$

and so $P(\omega) - 1 = 0$.

(b) We start by showing that $F(T) := 1 + T + T^2 + \cdots + T^{\ell-1}$ is the minimal polynomial of ω over $\mathbb{Q}$. Writing $F(T)$ as $\frac{T^\ell-1}{T-1}$ makes it clear that $F(\omega) = 0$, and so it is enough to show that $F(T)$ is irreducible over $\mathbb{Q}$. It turns out to be easier to prove the equivalent result that $F(T + 1)$ is irreducible over $\mathbb{Q}$: Start by noticing that, modulo ℓ,

$$F(T+1) = \frac{(T+1)^\ell - 1}{T} \equiv \frac{(T^\ell + 1) - 1}{T} = T^{\ell-1},$$

so that ℓ divides all of the non-leading coefficients of $F(T + 1)$. Then plug in $T = 0$ to see that the constant term of $F(T + 1)$ is $F(1) = \ell$, which is not divisible by ℓ^2. Eisenstein's criterion does the rest.

Since $P(\omega) - 1 = 0$ and F is the minimal polynomial of ω over $\mathbb{Q}$, it must be that $P(T) - 1 = F(T)Q(T)$ for some $Q(T) \in \mathbb{Q}[T]$. Since $P(T) - 1 \in \mathbb{Z}[T]$, and $F(T)$ is monic with integer coefficients, polynomial long division (or Gauss's polynomial lemma) shows that in fact $Q(T) \in \mathbb{Z}[T]$.

Letting $T = 1$ in (b), we find that $P(1) = 1 + \ell Q(1)$, so that $P(1) \equiv 1 \pmod{\ell}$. On the other hand, from the definition of $P(T)$,

$$P(1) = \left(\prod_{\substack{0<r<\ell \\ (\frac{r}{\ell})=1}} (1 + T^r + \cdots + T^{(n-1)r}) \right) \Bigg|_{T=1}$$

$$= n^{\#\{0<r<\ell:\,(\frac{r}{\ell})=1\}} = n^{\frac{\ell-1}{2}} \equiv -1 \pmod{\ell},$$

where the congruence in the last step comes from Euler's criterion and our choice of n as a nonsquare mod ℓ. But ℓ is odd, and so $1 \not\equiv -1 \pmod{\ell}$. Contradiction!

14.149 The number of reduced fractions of denominator n that belong to $[0, 1]$ is 2 when $n = 1$, and $\phi(n)$ when $n > 1$. Thus, $\Phi(N) = 1 + \sum_{n\leq N} \phi(n)$. So it is enough to prove that $\sum_{n\leq N} \phi(n)$ belongs to each congruence class mod 3 for infinitely many N.

Let χ be the (unique) nontrivial Dirichlet character modulo 3, which appeared already in Problem 8.83. Notice that χ is integer-valued, with $\chi(m) \equiv m \pmod 3$ for all m. Hence, setting $A(t) = \sum_{n\leq t} \chi(\phi(n))$,

$$\sum_{n \leq t} \phi(n) \equiv A(t) \pmod{3} \tag{B.52}$$

for all $t > 0$.

To complete the proof it will suffice to show that $A(t)$ is an unbounded function of t. Why is this enough? $A(t)$ is a step function, with jumps of ± 1 at positive integers that are not multiples of 3. So in order for $A(t)$ to take on a value m, it must take on all integer values between 0 and m. Thus, the only way $A(t)$ can be unbounded is if $A(N)$ takes on either every positive integer value, or every negative integer value, as N ranges over $\mathbb{Z}^+$. Either way, (B.52) implies that $\sum_{n \leq N} \phi(n)$ hits each congruence class mod 3 for infinitely many N.

Suppose for a contradiction that $A(t)$ is bounded, say $|A(t)| \leq K$ for all t. Borrowing ideas from the solution to Problem 13.131, we then find that $L(s) := \sum_{n=1}^{\infty} \frac{\chi(\phi(n))}{n^s}$ converges, for all $s > 0$, to $s \int_1^{\infty} A(t) t^{-s-1} \, dt$. Hence, for all $s > 0$,

$$|L(s)| \leq Ks \int_1^{\infty} t^{-s-1} \, dt = K.$$

But we will show in the next paragraph that $\lim_{s \downarrow 1} L(s) = \infty$.

The function $\chi(\phi(n))$ is multiplicative, and $\sum_{n=1}^{\infty} \frac{\chi(\phi(n))}{n^s}$ converges absolutely when $s > 1$. So by the remark following the solution to Problem 3.25,

$$L(s) = \prod_p \left(1 + \frac{\chi(p-1)}{p^s} + \frac{\chi(p(p-1))}{p^{2s}} + \frac{\chi(p^2(p-1))}{p^{3s}} + \cdots\right)$$

$$= \left(1 - \frac{1}{3^s}\right) \prod_{p \equiv 2 \,(\text{mod } 3)} \left(1 + \frac{1}{p^s} - \frac{1}{p^{2s}} + \frac{1}{p^{3s}} - \frac{1}{p^{4s}} + \cdots\right)$$

whenever $s > 1$. It follows that for each $s > 1$,

$$L(s) \geq \frac{2}{3} \prod_{p \equiv 2 \,(\text{mod } 3)} \left(1 + \frac{1}{p^s} - \frac{1}{p^{2s}}\right)$$

$$\geq \frac{2}{3} \left(1 + \sum_{p \equiv 2 \,(\text{mod } 3)} \left(\frac{1}{p^s} - \frac{1}{p^{2s}}\right)\right)$$

$$\geq \frac{2}{3} \left(1 - \sum_p \frac{1}{p^2} + \sum_{p \equiv 2 \,(\text{mod } 3)} \frac{1}{p^s}\right)$$

$$\geq \frac{2}{3} \sum_{p \equiv 2 \,(\text{mod } 3)} \frac{1}{p^s}.$$

(We use here that $\sum_p \frac{1}{p^2} \leq \sum_{n \geq 2} \frac{1}{n^2} = \zeta(2) - 1 \leq 2 - 1 = 1$.) The final sum on p blows up as $s \downarrow 1$, as follows from Problem 8.83(d). Hence, $L(s)$ also blows up as $s \downarrow 1$.

References

1. David C. Ullrich, *Complex made simple*. Graduate Studies in Mathematics, vol. 97. American Mathematical Society, Providence, RI, 2008.
2. James Victor Uspensky and Max A. Heaslet, *Elementary Number Theory*. McGraw-Hill Book Company, Inc., New York, 1939.

Solutions to Special Step C

15.150 The solution offered for Problem 14.140(a) applies almost verbatim: Replace $\mathbb{U}_\ell$ with G and $\ell - 1$ with $|G|$.

15.151 Since G is a finite group, $g^{|G|} \in H$, so that there is a minimal positive integer n with $g^n \in H$.

(a) Recall that in a finite group, a nonempty subset closed under the group operation is necessarily a subgroup.

Let $a, b \in H'$. Suppose that $a \in g^i H$ while $b \in g^j H$, with $0 \le i, j < n$. Writing $i + j = nq + r$, where $q, r \in \mathbb{Z}$ and $0 \le r < n$, we conclude that $ab \in g^{i+j} H = g^r g^{nq} H = g^r H \subseteq H'$. Thus, H' is a subgroup of G.

Since every coset of H has $|H|$ elements, in order to prove that $|H'| = n|H|$ it is enough to show that $g^i H$ and $g^j H$ are disjoint if $0 \le i < j \le n$. If not, then $g^i h_1 = g^j h_2$ for some $h_1, h_2 \in H$, forcing $g^{j-i} = h_2^{-1} h_1 \in H$, contrary to the minimality of n.

(b) Let χ be a character of H, and suppose that $\tilde{\chi}$ is an extension of χ to H'. Then writing h_0 for g^n,

$$\tilde{\chi}(g)^n = \tilde{\chi}(g^n) = \tilde{\chi}(h_0) = \chi(h_0),$$

so that $\tilde{\chi}(g)$ must be one the n nth roots of the nonzero complex number $\chi(h_0)$. It is clear from the definition of H' that the value of $\tilde{\chi}(g)$ determines the extension, and so χ has at most n extensions to H'. To see that χ has exactly n extensions, it is enough to produce, for any prescribed complex nth root ω of $\chi(h_0)$, an extension $\tilde{\chi}$ of χ with $\tilde{\chi}(g) = \omega$. For this, we simply define $\tilde{\chi}$ on H' by setting

$$\tilde{\chi}(g^i h) = \omega^i \chi(h) \qquad \text{for each } i \in \mathbb{Z}, h \in H.$$

© Springer Nature Switzerland AG 2021
P. Pollack, A. Singha Roy, *Steps into Analytic Number Theory*, Problem Books in Mathematics, https://doi.org/10.1007/978-3-030-65077-3_30

The only tricky bit is seeing that $\tilde{\chi}$ is well-defined; once this is verified, it is straightforward to check that $\tilde{\chi}$ is a character of H', and clearly $\tilde{\chi}(g) = \omega$.

To see that $\tilde{\chi}$ is well-defined, assume that $g^{i_1}h_1 = g^{i_2}h_2$, where $i_1, i_2 \in \mathbb{Z}$ and $h_1, h_2 \in H$. We must show that $\omega^{i_1}\chi(h_1) = \omega^{i_2}\chi(h_2)$. We start by noting that

$$g^{i_1-i_2} = h_2 h_1^{-1} \in H.$$

Writing $i_1 - i_2 = nq + r$, with $q, r \in \mathbb{Z}$ and $0 \le r < n$, we find (by an argument seen already in part a) that $g^{i_1-i_2}H = g^r H$. Since $g^{i_1-i_2} \in H$, it follows that $g^r \in H$, and so (by the minimality of n) $r = 0$. Hence, $i_1 - i_2 = nq$, so that

$$g^{i_i-i_2} = g^{nq} = h_0^q, \quad \text{and}$$

$$\omega^{i_i-i_2} = \omega^{nq} = (\omega^n)^q = \chi(h_0)^q = \chi(h_0^q).$$

Consequently,

$$\begin{aligned}
\omega^{i_1}\chi(h_1) &= \omega^{i_2}\omega^{i_1-i_2}\chi(h_1) \\
&= \omega^{i_2}\chi(h_0^q h_1) \\
&= \omega^{i_2}\chi((g^{i_1-i_2})h_1) \\
&= \omega^{i_2}\chi((h_2 h_1^{-1})h_1) \\
&= \omega^{i_2}\chi(h_2).
\end{aligned}$$

15.152 The group H' constructed from H and g in Problem 15.151(a) is precisely the subgroup of G generated by H and g. Moreover, the integer n appearing in that construction is precisely the index of H in H'. So we can rephrase the conclusion of Problem 15.151 as follows: If H is a subgroup of G, $g \in G$, and H' is the subgroup of G generated by H and g, then each character of H extends in $[H' : H]$ ways to a character of H'. We assumed in Problem 15.151 that $g \notin H$, but if $g \in H$ the statement is trivial.

Now let $g_1, g_2, \ldots, g_m$ be any finite generating set for G. Let H_i be the subgroup generated by $g_1, \ldots, g_i$, for each integer $0 \le i \le m$.[1] When $i = 0$, the subgroup H_i consists only of the identity, and the number of characters of H_i is $1 = |H_i|$. Suppose now that $i < m$ and that H_i is known to have exactly $|H_i|$ characters. Restricting a character of H_{i+1} to H_i yields a character of H_i. Looking at it from the other end, each character of H_{i+1} can be obtained by extending a character of

---

[1] When we speak of the "subgroup of G generated by a set S", we mean the minimal subgroup of G containing S, with respect to inclusion. In particular, the subgroup generated by $\emptyset$ is the trivial subgroup of G.

H_i. But each character of H_i extends in $[H_{i+1} : H_i]$ ways to a character of H_{i+1}. Hence, the number of characters of H_{i+1} is

$$|H_i| \cdot [H_{i+1} : H_i] = |H_{i+1}|.$$

By induction, each H_i has exactly $|H_i|$ characters, and in particular $G = H_m$ has exactly $|G|$ characters.

15.153 Let H be the subgroup generated by g, and choose $g_1, \ldots, g_m \in G$ such that H and $g_1, \ldots, g_m$ generate G. For each $i = 0, 1, 2, \ldots, m$, let H_i be the subgroup of G generated by H and $g_1, \ldots, g_i$. A straightforward induction, using Problem 15.151, shows that each character of H extends to a character of H_i, for all $i = 0, 1, 2, \ldots, m$. In particular, each character of H extends to $G = H_m$. Thus, it is enough to prove that there is a character χ of H with $\chi(g) \neq 1$. This is easy: Since H is generated by g, the characters of H are determined by how they act on g. Thus, at most one character χ of H can satisfy $\chi(g) = 1$, while there are $|H| > 1$ characters of H in total. (We use only in the last step that g is not the identity, to guarantee that $|H| > 1$.)

15.154 After the last exercise, we can follow the proof for the corresponding claim given in the solution to Problem 14.138.

15.155 Let $A_\chi(t) = \sum_{n \leq t} \chi(n)$. The sum of $\chi(n)$ vanishes when n runs over any block of m consecutive integers, and so $|A_\chi(t)| < m$ for all t. For all real $s > 0$,

$$\sum_{n \leq x} \frac{\chi(n)}{n^s} = x^{-s} A_\chi(x) + s \int_1^x \frac{A_\chi(t)}{t^{s+1}} \, dt, \qquad (C.53)$$

while

$$L(s, \chi) = s \int_1^\infty \frac{A_\chi(t)}{t^{s+1}} \, dt.$$

(Cf. the solutions to Problems 13.131 and 14.140.) Thus,

$$\left| L(s, \chi) - \sum_{n \leq x} \frac{\chi(n)}{n^s} \right| = \left| s \int_x^\infty \frac{A_\chi(t)}{t^{s+1}} \, dt - x^{-s} A_\chi(x) \right|.$$

To conclude, observe that when $s = 1$, the right-hand side is bounded above by $m \int_x^\infty \frac{dt}{t^2} + mx^{-1} = 2mx^{-1}$.

15.156 This follows by taking $s = 1$ in (C.53), since $x^{-1} A_\chi(x) = O_m(x^{-1})$.

15.157 By Problems 15.155 and 15.156, $L(1, \chi) = \int_1^x (\sum_{n \leq t} \chi(n)) t^{-2} \, dt + O_m(x^{-1})$, for every $x \geq 1$. Hence, for each $x \geq 1$,

$$R(x) - xL(1, \chi) = \int_1^x \left(\sum_{n \le t} \chi(n) \right) \left\lfloor \frac{x}{t} \right\rfloor \frac{dt}{t} - x \int_1^x \left(\sum_{n \le t} \chi(n) \right) t^{-2} dt + O_m(1)$$

$$= \int_1^x \left(\sum_{n \le t} \chi(n) \right) \left(\left\lfloor \frac{x}{t} \right\rfloor \frac{1}{t} - \frac{x}{t^2} \right) dt + O_m(1).$$

Noting that $-\frac{1}{t} \le \lfloor \frac{x}{t} \rfloor \frac{1}{t} - \frac{x}{t^2} \le 0$, and recalling that $| \sum_{n \le t} \chi(n) | < m$, we see that the remaining integral satisfies

$$\left| \int_1^x \left(\sum_{n \le t} \chi(n) \right) \left(\left\lfloor \frac{x}{t} \right\rfloor \frac{1}{t} - \frac{x}{t^2} \right) dt \right| \le m \int_1^x \frac{1}{t} dt = m \log x.$$

Hence, $R(x) - xL(1, \chi) = O_m(\log x) + O_m(1) = O_m(\log (ex))$, for $x \ge 1$.

15.158 Using 1_C for the indicator function of the condition C,

$$R(x) = \int_1^x \left(\sum_{nm \le x} \chi(n) 1_{n \le t} 1_{m \le x/t} \right) \frac{dt}{t}$$

$$= \sum_{nm \le x} \chi(n) \int_1^x 1_{n \le t} 1_{m \le x/t} \frac{dt}{t}$$

$$= \sum_{nm \le x} \chi(n) \int_n^{x/m} \frac{dt}{t},$$

and so we have the first equality claimed in the problem statement. To obtain the second, rewrite

$$\sum_{nm \le x} \chi(n) \int_n^{x/m} \frac{dt}{t} = \sum_{nm \le x} \chi(n) \log \frac{x}{nm}$$

$$= \sum_{k \le x} \left(\sum_{nm = k} \chi(n) \right) \log \frac{x}{k} = \sum_{k \le x} r(k) \log \frac{x}{k}.$$

15.159 Write $n = p_1^{e_1} \cdots p_k^{e_k}$. The divisors of n are precisely the numbers $p_1^{f_1} \cdots p_k^{f_k}$, with each $0 \le f_i \le e_i$. Thus,

$$\sum_{d \mid n} \chi(d) = \prod_{i=1}^k \left(1 + \chi(p_i) + \chi(p_i^2) + \cdots + \chi(p_i^{e_i}) \right),$$

as one sees upon multiplying out the right-hand side. Since the left side here is $r(n)$, while the ith factor on the right is $r(p_i^{e_i})$, it follows immediately that $r(n)$ is multiplicative.

With this out of the way, to establish the remaining two assertions, it is enough to show that $r(p^e) \geq 0$ for every prime power p^e, and that $r(p^e) \geq 1$ when e is even. In fact,

$$r(p^e) = 1 + \chi(p) + \chi(p)^2 + \cdots + \chi(p)^e = \begin{cases} 1 & \text{if } \chi(p) = 0, \\ e+1 & \text{if } \chi(p) = 1, \\ 1 & \text{if } \chi(p) = -1 \text{ and } 2 \mid e, \\ 0 & \text{if } \chi(p) = -1 \text{ and } 2 \nmid e. \end{cases}$$

15.160 From Problems 15.159 and 15.158,

$$R(x) = \sum_{k \leq x} r(k) \log \frac{x}{k} \geq \sum_{m^2 \leq x} \log \frac{x}{m^2} = \lfloor \sqrt{x} \rfloor \log x - 2 \log \lfloor \sqrt{x} \rfloor !$$

$$= \sqrt{x} \log x - 2(\sqrt{x} \log \sqrt{x} - \sqrt{x}) + O(\log (ex)) = 2\sqrt{x} + O(\log (ex)).$$

(Here we estimated $\log \lfloor \sqrt{x} \rfloor !$ using Problem 2.12.) This estimate certainly implies that $R(x) > x^{1/2}$ for all large enough x.

15.161 Suppose, by way of contradiction, that $L(1, \chi) = 0$. By Problem 15.157, $R(x) = O_m(\log (ex))$. This contradicts the result of Problem 15.160.

Remark. Showing that $L(1, \chi) \neq 0$ for real, nontrivial χ is the most difficult step in the proof of Dirichlet's theorem. Our argument for this feels (and is) rather ad hoc. That proof was chosen because it is relatively simple, with minimal prerequisites, but it is far from natural. The most satisfying proof of nonvanishing is surely the one found by Dirichlet himself, connecting the values of $L(1, \chi)$ to the arithmetic of degree 2 extensions of $\mathbb{Q}$. The proof is beyond our present scope, but the result itself is so remarkable that it merits a brief description.

Let χ be a Dirichlet character mod m. Let m_1 be any multiple of m, and let χ_0 be the trivial character mod m_1. Then $\tilde{\chi} := \chi_0 \chi$ is a Dirichlet character modulo m_1, referred to as the character mod m' "induced" by χ. Moreover,

$$L(1, \tilde{\chi}) = L(1, \chi) \prod_{p \mid m_1} \left(1 - \frac{\chi(p)}{p}\right), \tag{C.54}$$

so that $L(1, \tilde{\chi})$ is nonzero precisely when $L(1, \chi)$ is nonzero.

A Dirichlet character mod m is said to be "primitive" if it cannot be induced by a character whose modulus is a proper divisor of m. Each Dirichlet character is induced by a primitive character, and nontrivial real characters are induced by nontrivial real primitive characters. So by (C.54), in order to show the nonvanishing of $L(1, \chi)$ for nontrivial real χ, it is enough to tackle the case when χ is primitive.

Let χ be a real primitive character to the modulus m. There is a uniquely determined squarefree integer d with the property that

$$\chi(p) = \left(\frac{d}{p}\right) \quad \text{for all sufficiently large primes } p. \tag{C.55}$$

(One can show that $d = \pm m$ or $d = \pm m/4$.) A rough statement of Dirichlet's result (his "quadratic class number formula") is that

$$L(1, \chi) \approx \begin{cases} h_d \frac{\pi}{\sqrt{|d|}} & \text{when } d < 0, \\ h_d \frac{\log \epsilon}{\sqrt{d}} & \text{when } d > 0. \end{cases} \tag{C.56}$$

Our notation here requires explanation. The number h_d is the "class number" of the ring of integers of $\mathbb{Q}(\sqrt{d})$. This is a positive integer whose size measures failure of unique factorization in that ring. The number ϵ_d, which appears in the formula only when $d > 0$, is the "fundamental unit" of the ring of integers of $\mathbb{Q}(\sqrt{d})$. This is the generator larger than 1 of the group of positive units of that ring. Finally, the symbol "$\approx$" means that the quotient of the left and right sides is a positive rational number (given explicitly by Dirichlet) for which there are only finitely many possibilities. All of the quantities appearing on the right of (C.56) are nonzero, and so the nonvanishing of $L(1, \chi)$ follows immediately!

We content ourselves with one example. Let χ be the character mod 8 labeled as χ_2 in Special Step A. Then

$$L(1, \chi) = 1 - \frac{1}{3} - \frac{1}{5} + \frac{1}{7} + \frac{1}{9} - \frac{1}{11} - \frac{1}{13} + \frac{1}{15} + \dots$$

$$= \int_0^1 (t^0 - t^2 - t^4 + t^6 + t^8 - t^{10} - t^{12} + t^{14} + \dots) \, dt$$

$$= \int_0^1 \frac{1 - t^2 - t^4 + t^6}{1 - t^8} \, dt.$$

With some elbow grease, the last integral succumbs to the method of partial fractions, and we find that

$$L(1, \chi) = \frac{\log(\sqrt{2} + 1)}{\sqrt{2}}. \tag{C.57}$$

How does this match up with (C.56)? Our χ is a primitive character mod 8, and $d = 2$ in (C.55). The ring of integers of $\mathbb{Q}(\sqrt{2})$ is $\mathbb{Z}[\sqrt{2}]$. That ring is a unique factorization domain, so that $h_2 = 1$, and the fundamental unit in the ring is $\epsilon_2 = \sqrt{2} + 1$. So (C.57) matches up exactly with the right-hand side of (C.56) (that is, the correction factor implied by "$\approx$" is not necessary in this case).

There were several bold, unproved assertions in the last few paragraphs. For the missing details, the reader can consult any (or all) of [1, 4], and [5].

15.162 Suppose that $(m, n) \neq (2, 1)$. Then $mn \geq 3$, and we can fix a prime q dividing $mn - 1$. Then m and n are inverses modulo q, and so both are certainly coprime to q.

It is enough to prove that q divides $m^p - n$ for infinitely many distinct primes p. Working mod q,

$$n^p(m^p - n) \equiv (mn)^p - n^{p+1} \equiv 1 - n^{p+1}.$$

By Dirichlet's theorem, there are infinitely many p with $p \equiv -1 \pmod{q - 1}$. For all of these primes, $n^{p+1} \equiv 1 \pmod{q}$, and so

$$n^p(m^p - n) \equiv 1 - n^{p+1} \equiv 0 \pmod{q}.$$

Thus, $q \mid n^p(m^p - n)$. But $q \nmid n$, and so $q \mid m^p - n$.

15.163 The difference in question is precisely $\sum_{p \leq x} \chi(p)$, where χ is the nontrivial Dirichlet character mod 4.

Suppose that for some constant K and all real numbers x,

$$\left| \sum_{p \leq x} \chi(p) \right| \leq K.$$

The same partial summation argument we used to analyze $L(s, \chi)$ will then show that the function $\sum_p \frac{\chi(p)}{p^s}$ is analytic for $\Re(s) > 0$ and that, in this same half-plane,

$$\sum_p \frac{\chi(p)}{p^s} = s \int_1^\infty \left(\sum_{p \leq t} \chi(p) \right) t^{-s-1} \, dt.$$

In particular, when s is real and positive,

$$\left| \sum_p \frac{\chi(p)}{p^s} \right| \leq K s \int_1^\infty t^{-s-1} \, dt = K.$$

Recall that when $\Re(s) > 1$,

$$\text{Log } L(s, \chi) = \sum_p \frac{\chi(p)}{p^s} + \sum_p \frac{\chi(p^2)}{2p^{2s}} + \sum_p \sum_{k>3} \frac{\chi(p^k)}{kp^{ks}}$$

$$= A(s) + B(s) + C(s), \tag{C.58}$$

say. It is not difficult to show that the series defining $B(s)$ converges uniformly on compact subsets of $\Re(s) > \frac{1}{2}$, so that $B(s)$ is analytic for $\Re(s) > \frac{1}{2}$. Similar arguments show that $C(s)$ is analytic for $\Re(s) > \frac{1}{3}$ and bounded on every closed half-plane $\Re(s) \geq \frac{1}{3} + \delta$, with $\delta > 0$.

Exponentiating (C.58),

$$L(s, \chi) = \exp(A(s) + B(s) + C(s))$$

whenever $\Re(s) > 1$. By the principle of analytic continuation, this identity for $L(s, \chi)$ remains valid in the half-plane $\Re(s) > \frac{1}{2}$. Now we restrict to real s and consider what happens as $s \downarrow \frac{1}{2}$. We have seen already that $A(s)$ and $C(s)$ remain bounded. On the other hand, as $s \downarrow \frac{1}{2}$,

$$B(s) = \sum_p \frac{\chi(p^2)}{2p^{2s}} = \frac{1}{2} \sum_{p \text{ odd}} \frac{1}{p^{2s}} \to \infty.$$

Hence, $L(s, \chi)$ blows up. But this is absurd: As $s \downarrow \frac{1}{2}$, the (continuous) function $L(s, \chi)$ tends to the finite value $L(\frac{1}{2}, \chi)$.

This proof is taken from a `mathoverflow` answer by the user `lucia` [3].

Remark. Thus, if we tally primes up to higher and higher stopping points, from time to time we will see either many more 1 mod 4 primes than 3 mod 4 primes, or vice versa. In fact, both scenarios play out infinitely often: $\sum_{p \leq x} \chi(p)$ takes on arbitrarily large positive values as well as arbitrarily large negative values. This was proved by Littlewood in 1914 (in a precise quantitative form). On the other hand, $\sum_{p \leq x} \chi(p)$ *prefers* to be negative: Rubinstein and Sarnak have shown, conditional on plausible conjectures about the complex zeros of $\zeta(s)$ and $L(s, \chi)$, that $\sum_{p \leq x} \chi(p) < 0$ for about 99.59% of positive integers x.[2] For an overview of the techniques used to study such problems, we refer the reader to the survey article of Granville and Martin [2].

References

1. Harvey Cohn, *Advanced number theory*. Reprint of *A second course in number theory*, 1962. Dover Books on Advanced Mathematics. Dover Publications, Inc., New York, 1980.
2. Andrew Granville and Greg Martin, *Prime number races*. Amer. Math. Monthly 113 (2006), 1–33.
3. Lucia (https://mathoverflow.net/users/38624/lucia), The prime numbers modulo k, are not periodic, URL (version: 2014-05-28): https://mathoverflow.net/q/168465
4. Paul Pollack, *A conversational introduction to algebraic number theory*. Student Mathematical Library, vol. 84. American Mathematical Society, Providence, RI, 2017.
5. Don B. Zagier, *Zetafunktionen und quadratische Körper*. Hochschultext. Springer-Verlag, Berlin-New York, 1981.

---

[2]Here the percentage claim is to be interpreted in the sense of logarithmic density. That is: $\frac{1}{\log X} \sum'_{x \leq X} \frac{1}{x}$ tends (as $X \to \infty$) to a limit that is ≈ 0.9959. The ′ means that we sum only over integers $x \leq X$ for which there are more 3 mod 4 primes up to x than 1 mod 4 primes.

Epilogue

> Theorems are fun especially when you
> are the prover, but then the pleasure
> fades. What keeps us going are the
> unsolved problems.
>
> Carl Pomerance

By necessity, problem sets that attempt to introduce an area of modern research
will always paint an incomplete picture. Below we briefly discuss the state of our
knowledge (and ignorance) concerning some of the main topics we explored in these
problem sets.

Distribution of squares and nonsquares modulo p. It was conjectured by Vinogradov around 1920 that for each $\epsilon > 0$, the least positive nonsquare modulo
p is $< p^\epsilon$, for all sufficiently large primes p. Despite being an intense focus of
research over the past century, Vinogradov's conjecture remains unsolved.

Some progress has been made. Linnik proved in 1940 that counterexamples
to Vinogradov's conjecture are rare: For all $x \geq 3$, there are $O_\epsilon(\log \log x)$
primes $p \leq x$ with least nonsquare $> p^\epsilon$. In fact, if we replace $> p^\epsilon$ with $> x^\epsilon$,
then the count of such $p \leq x$ shrinks to $O_\epsilon(1)$. And around 1960, Burgess
proved that for each $\epsilon > 0$, the least positive nonsquare modulo p is $< p^{\frac{1}{4\sqrt{e}}+\epsilon}$,
once p is large enough. (The exponent here is half of that of Problem 11.109.)
His main tool was an estimate for Legendre symbol sums sharper than that of
Problem 6.55. Replacing $\frac{1}{4\sqrt{e}}$ by any smaller number would be a major advance.

Arithmetic functions and the anatomy of integers. The results of Step #11 on
the typical sizes of $\omega(n)$ and $\Omega(n)$ are due to Hardy and Ramanujan (1917).
Our (simpler) approach follows Turán (1934). Hardy and Ramanujan's proofs
were based on the following inequality, which has since found many other
applications: For certain constants K and K', and all real $x \geq 2$ and integers
$k \geq 1$,

© Springer Nature Switzerland AG 2021
P. Pollack, A. Singha Roy, *Steps into Analytic Number Theory*, Problem Books in
Mathematics, https://doi.org/10.1007/978-3-030-65077-3_31

$$\sum_{\substack{n \le x \\ \omega(n)=k}} 1 \le K \frac{x}{\log x} \frac{(\log \log x + K')^{k-1}}{(k-1)!}. \qquad (*)$$

That this upper bound holds when $k = 1$ (for an appropriate K, and any K') is a straightforward consequence of Problem 8.78. Writing $\pi_k(x)$ for the left-hand side of (*), one observes that $k\pi_{k+1}(x) \le \sum_{p^e \le \sqrt{x}} \pi_k(x/p^e)$. (Can you see why?) The proof is completed by induction on k, once one shows that for a suitably chosen K',

$$\sum_{p^e \le \sqrt{x}} \frac{1}{p^e \log(x/p^e)} \le \frac{\log \log x + K'}{\log x}.$$

You are invited to fill in the details!

Later, Erdős and Kac (1940) established the remarkable fact that $\omega(n)$ and $\Omega(n)$, for $n \le x$, are (as $x \to \infty$) asymptotically normally distributed with mean and variance $\log \log x$. Precisely: Let $f \in \{\omega, \Omega\}$. Then for each $z \in \mathbb{R}$,

$$\lim_{x \to \infty} \frac{1}{x} \#\{n \le x : \frac{\omega(n) - \log \log x}{\sqrt{\log \log x}} \le z\} = \frac{1}{\sqrt{2\pi}} \int_{-\infty}^{z} e^{-\frac{1}{2}t^2} \, dt.$$

(Notice that this subsumes the result that $\omega(n)$, $\Omega(n)$ are typically $\approx \log \log x$ for integers $n \le x$.) This result of Erdős and Kac was a major impetus for the development of the subject that has come to be known as "probabilistic number theory". See Fig. 17.1.

There is no shortage of open problems concerning the behavior of arithmetic functions. To take an example, let $M(N)$ denote the number of elements in the

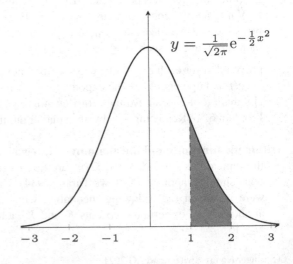

Fig. 17.1 By the Erdős–Kac theorem, the limiting proportion of $n \le x$ (as $x \to \infty$) with $1 < \frac{\omega(n) - \log \log x}{\sqrt{\log \log x}} \le 2$ is the area of the shaded region

$$y = \frac{1}{\sqrt{2\pi}} e^{-\frac{1}{2}x^2}$$

Table 17.1 Comparison of $\pi(x)$ and Li(x) (rounded to the nearest integer). While Li$(x) > \pi(x)$ for all x shown, Littlewood showed that there are arbitrarily large values of x where the reverse inequality holds

x	$\pi(x)$	Li(x)	Li$(x) - \pi(x)$
10^3	168	177	9
10^4	1229	1245	16
10^5	9592	9629	37
10^6	78,498	78,627	129
10^7	664,579	664,917	338
10^8	5,761,455	5,762,208	753
10^9	50,847,534	50,849,234	1700
10^{10}	455,052,511	455,055,614	3103
10^{11}	4,118,054,813	4,118,066,400	11,587
10^{12}	37,607,912,018	37,607,950,280	38,262
10^{13}	346,065,536,839	346,065,645,809	108,970

$N \times N$ multiplication table. You showed in Problem 12.120 that $M(N)/N^2 \to 0$ as $N \to \infty$. It was proved by Kevin Ford (2008) that $M(N)$ lies between two positive constant multiples of

$$M_\approx(N) := \frac{N^2}{(\log N)^\delta (\log\log N)^{3/2}}, \quad \text{where } \delta = 1 - \frac{1 + \log\log 2}{\log 2} \quad (\approx 0.086),$$

once N is large enough. Question: Does $M(N)/M_\approx(N)$ tend to a limit as $N \to \infty$? Extensive computations on this problem have been carried out recently by Brent, Pomerance, Purdum, and Webster, but it remains unclear what the numbers are trying to tell us.

The distribution of prime numbers. We alluded in Step #12 to the Prime Number Theorem, which we formulated as a result comparing $\pi(x)$ to $x/\log x$. But it was recognized already by Gauss that $\pi(x)$ ought to be compared not to $x/\log x$ but to the "logarithmic integral" Li$(x) := \int_2^x \frac{dt}{\log t}$. Since $\frac{x/\log x}{\text{Li}(x)} \to 1$ as $x \to \infty$, the Prime Number Theorem—being an assertion about a limit of ratios—can be stated either way. However, as far as absolute numerical difference with $\pi(x)$ is concerned, the logarithmic integral is a much closer approximation than $x/\log x$ (Table 17.1). How close?

Conjecture $|\pi(x) - \text{Li}(x)| < \sqrt{x}\log x$, for all $x \geq 3$.

Actually, this Conjecture is no more (and no less) than the infamous "Riemann Hypothesis"! To see the relation with the usual formulation of RH in terms of zeros of (the analytically continued version of) $\zeta(s)$, look up "Riemann's explicit formula" (for example, in [1]).

Prime numbers in arithmetic progressions. Dirichlet's methods imply that for every positive integer m, the prime numbers are equidistributed over the coprime residue classes modulo m, in the following sense: For every integer a coprime to m,

$$\lim_{x \to \infty} \frac{\sum_{p \leq x, \; p \equiv a \pmod m} 1/p}{\sum_{p \leq x} 1/p} = \frac{1}{\phi(m)}.$$

It is tempting to ask if the same limiting relation holds when the summand $\frac{1}{p}$ is replaced by 1, as that would be a significantly more natural way of measuring equidistribution. In other words, if we set

$$\pi(x; m, a) = \#\{p \leq x : p \equiv a \pmod m\},$$

does $\frac{\pi(x;m,a)}{\pi(x)} \to \frac{1}{\phi(m)}$, as $x \to \infty$ (assuming a and m are coprime)? YES! This is called the Prime Number Theorem for Arithmetic Progressions, and it can be proved by the same techniques used to establish the ordinary Prime Number Theorem.

The PNT for APs is not the end of the story. If we *fix* m, that result tells us that the primes $\leq x$ equidistribute themselves over the coprime progressions mod m once x is large enough compared to m. But how long do we have to wait for equidistribution to "kick in"? It is conjectured that $x > m^{1+\epsilon}$ suffices, but known results do not guarantee equidistribution unless $x > \exp(m^\epsilon)$.

Combinatorial methods. In Step #9, you finished the proof of Brun's theorem: $\sum_p \frac{1}{p} < \infty$, where p runs over the primes for which $p + 2$ is also prime. Brun's argument (which we followed) was the first important example (1919) of what is called a "sieve method" in number theory. Sieve methods have seen extensive development, and they continue to be the source of the sharpest results to date on the twin prime conjecture. We mention two: (a) (Chen, 1966) There are infinitely many primes p for which $p + 2$ is either prime or a product of two primes (counting multiplicity). (b) (D.H.J. Polymath, 2014, building on ideas of Zhang, Maynard, Tao) There are infinitely pairs of primes that differ by no more than 246.

Reference

1. Harold Davenport, *Multiplicative Number Theory* (3rd ed.), 2000: Springer, New York.

Suggestions for Further Reading

1. Tom M. Apostol, *Introduction to Analytic Number Theory*, 2010: Springer, New York.
2. Alina Cojocaru and M. Ram Murty, *An Introduction to Sieve Methods and their Applications*, 2005: Cambridge University Press, Cambridge.
3. Harold Davenport, *Multiplicative Number Theory* (3rd ed.), 2000: Springer,
4. Adolf J. Hildebrand, *Introduction to Analytic Number Theory: Lecture Notes*, 2013. Online resource: https://faculty.math.illinois.edu/~hildebr/ant/index.html
5. Edmund Hlawka, Johannes Schoißengeier, and Rudolf Taschner, *Geometric and Analytic Number Theory*, 1991: Springer-Verlag, Berlin.
6. Loo-Keng Hua, *Introduction to Number Theory*, 1987: Springer-Verlag, Berlin.
7. Edmund Landau, *Elementary Number Theory*, 1999: American Mathematical Society, Providence, RI.
8. Florian Luca and Jean-Marie De Koninck, *Analytic Number Theory: Exploring the Anatomy of Integers*, 2012: American Mathematical Society, Providence, RI.
9. Hugh L. Montgomery and Robert C. Vaughan, *Multiplicative Number Theory I. Classical Theory*, 2012: Cambridge University Press, Cambridge.
10. M. Ram Murty, *Problems in Analytic Number Theory* (2nd ed.), 2007: Springer, New York.
11. Melvyn B. Nathanson, *Additive Number Theory: The Classical Bases*, 1996: Springer, New York.
12. Paul Pollack, *Not Always Buried Deep: A Second Course in Elementary Number Theory*, 2009: American Mathematical Society, Providence, RI.
13. Harold N. Shapiro, *Introduction to the Theory of Numbers*, 2008: Dover, New York.
14. Gerald Tenenbaum, *Introduction to Analytic and Probabilistic Number Theory* (3rd ed.), 2015: American Mathematical Society, Providence, RI.

© Springer Nature Switzerland AG 2021   197
P. Pollack, A. Singha Roy, *Steps into Analytic Number Theory*, Problem Books in Mathematics, https://doi.org/10.1007/978-3-030-65077-3

Printed in the United States
by Baker & Taylor Publisher Services

Printed in the United States
by Baker & Taylor Publisher Services